SpringerBriefs in Applied Sciences and Technology

Nonlinear Circuits

Series editors

Luigi Fortuna, Catania, Italy
Guanrong Chen, Kowloon, Hong Kong SAR, P.R. China

W0193242

SpringerBriefs in Nonlinear Circuits promotes and expedites the dissemination of substantive new research results, state-of-the-art subject reviews and tutorial overviews in nonlinear circuits theory, design, and implementation with particular emphasis on innovative applications and devices. The subject focus is on nonlinear technology and nonlinear electronics engineering. These concise summaries of 50–125 pages will include cutting-edge research, analytical methods, advanced modelling techniques and practical applications. Coverage will extend to all theoretical and applied aspects of the field, including traditional nonlinear electronic circuit dynamics from modelling and design to their implementation. Topics include but are not limited to:

- nonlinear electronic circuits dynamics;
- oscillators;
- cellular nonlinear networks;
- arrays of nonlinear circuits;
- chaotic circuits;
- system bifurcation;
- chaos control;
- active use of chaos;
- nonlinear electronic devices;
- memristors;
- circuit for nonlinear signal processing;
- wave generation and shaping;
- nonlinear actuators;
- nonlinear sensors;
- power electronic circuits;
- nonlinear circuits in motion control;
- nonlinear active vibrations;
- educational experiences in nonlinear circuits;
- nonlinear materials for nonlinear circuits; and
- nonlinear electronic instrumentation.

Contributions to the series can be made by submitting a proposal to the responsible Springer contact, Oliver Jackson (oliver.jackson@springer.com) or one of the Academic Series Editors, Professor Luigi Fortuna (luigi.fortuna@dieei.unict.it) and Professor Guanrong Chen (eegchen@cityu.edu.hk).

Members of the Editorial Board:

More information about this series at http://www.springer.com/series/15574

Viet-Thanh Pham · Christos Volos
Tomasz Kapitaniak

Systems with Hidden Attractors

From Theory to Realization in Circuits

 Springer

Viet-Thanh Pham
School of Electronics
  and Telecommunications
Hanoi University of Science and Technology
Hanoi
Vietnam

Tomasz Kapitaniak
Faculty of Mechanical Engineering
Lodz University of Technology
Łódź
Poland

Christos Volos
Department of Physics
Aristotle University of Thessaloniki
Thessaloniki
Greece

ISSN 2191-530X   ISSN 2191-5318   (electronic)
SpringerBriefs in Applied Sciences and Technology
ISSN 2520-1433   ISSN 2520-1441   (electronic)
Nonlinear Circuits
ISBN 978-3-319-53720-7   ISBN 978-3-319-53721-4   (eBook)
DOI 10.1007/978-3-319-53721-4

Library of Congress Control Number: 2017931577

Printed on acid-free paper

This Springer imprint is published by Springer Nature
The registered company is Springer International Publishing AG
The registered company address is: Gewerbestrasse 11, 6330 Cham, Switzerland

Preface

Recently, Profs. Leonov and Kuznetsov have introduced a new classification of nonlinear dynamics in which they concentrated on two kinds of attractors: self-excited attractors and hidden attractors. A self-excited attractor has a basin of attraction that is excited from unstable equilibria. So from that point of view, most known nonlinear systems, such as Lorenz system, Rössler system, Chen system, or Sprott system, belong to chaotic systems with self-excited attractors. In contrast, a few unusual systems such as those with an infinite number of equilibrium points, with stable equilibria, or without equilibrium belong to systems with hidden attractors.

Studying systems with hidden attractors has become an attractive research direction because hidden attractors play an important role in theoretical problems and engineering applications. For example, hidden attractor can generate unexpected and potentially disastrous responses to perturbations in a structure like a bridge or an airplane wing. Therefore, it is useful for engineering students and researchers to have an overview of this new classification of attractors. This brief book is a concise reference in nonlinear systems with hidden attractors. Furthermore, emergent topics in circuit implementation of systems with hidden attractors are presented in this book. Also, this book can be used as a part of the bibliography in courses related to dynamical systems and their applications, nonlinear circuits, or oscillations in mechanical systems.

This book is organized as follows: Hidden attractor and its presence in nonlinear systems are presented briefly in Chap. 1. Systems with stable equilibrium, systems with an infinite number of equilibrium points, and systems without equilibrium are reported in Chaps. 2–4, respectively. In Chap. 5, we discuss synchronization of systems with hidden attractors. Chapter 6 introduces circuitry realizations of various systems with hidden attractors. Finally, conclusion remarks are drawn in Chap. 7.

We would like to acknowledge the helpful suggestions and discussions with Nikolay V. Kuznetsov and Mattia Frasca. Furthermore, we would like to thank our collaborators Sajad Jafari, Sundarapandian Vaidyanathan, Xiong Wang, Zhouchao

Wei, Fadhil Rahma, and Sifeu Takougang Kingni. We are thankful to Professor Luigi Fortuna who has encouraged us to prepare this book.

Finally, we would like to acknowledge the help and support of our families that include numerous "hidden attractors."

Hanoi, Vietnam   Viet-Thanh Pham
Łódź, Poland   Christos Volos
Thessaloniki, Greece   Tomasz Kapitaniak
January 2017

Contents

Chapter 1
Introduction

1.1 Self-Excited Attractors

During the initial period of the establishment of the theory of nonlinear oscillations that took place in the first half of the twentieth century [3, 134, 139], much attention was paid to the analysis and synthesis of oscillating systems, in which the problem of the existence of oscillations could be solved with relative ease. This approach was encouraged by the research on periodic oscillations in applied sciences, such as mechanics, electronics, chemistry, biology, and so on [135].

However, the first attempt on this direction was at the end of nineteenth century, with Rayleigh's works devoted to the study of string oscillations in musical instruments [120]. He first discovered that in a two-dimensional nonlinear dynamical system, undamped vibrations can be arisen without external periodic action. The Rayleigh system is described as:

$$\ddot{x} - \left(a - b\dot{x}^2\right)\dot{x} + x = 0 \tag{1.1}$$

By using $a = 1$ and $b = 0.1$, in the Rayleigh system (1.1), a limit cycle is produced as final state of the system. In Fig. 1.1, the produced limit cycle is illustrated for two different sets of initial conditions.

The extension of system (1.1) is a well-known van der Pol equation [114].

$$\ddot{x} + \mu \left(x^2 - 1\right)\dot{x} + x = 0 \tag{1.2}$$

This equation was originally proposed by the Dutch electrical engineer and physicist Balthasar van der Pol while he was working at Philips Labs. van der Pol found stable oscillations, which he subsequently called relaxation oscillations and are now known as a type of limit cycle in electrical circuits employing vacuum tubes. In Fig. 1.2, the produced limit cycle is illustrated for two different sets of initial conditions.

In 1951, Belousov first discovered oscillations in chemical reactions in a liquid phase [9]. Consider one of the Belousov–Zhabotinsky dynamic models:

© The Author(s) 2017
V.-T. Pham et al., *Systems with Hidden Attractors*,
SpringerBriefs in Nonlinear Circuits, DOI 10.1007/978-3-319-53721-4_1

Fig. 1.1 Phase portrait in the $x - \dot{x}$ plane for the Rayleigh system, for initial conditions $(x(0), \dot{x}(0)) = (0.5, 0.5)$ (*blue color*) and $(x(0), \dot{x}(0)) = (6, 6)$ (*red color*)

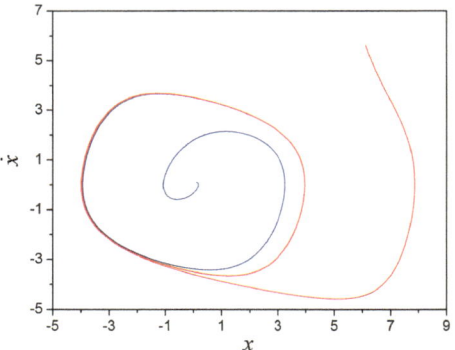

Fig. 1.2 Phase portrait in the $x - \dot{x}$ plane for the van der Pol system with $\mu = 2$, for initial conditions $(x(0), \dot{x}(0)) = (0.5, 0.5)$ [blue color] and $(x(0), \dot{x}(0)) = (6, 6)$ (*red color*)

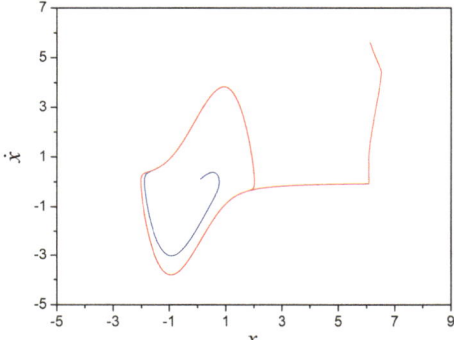

$$\begin{cases} \varepsilon \dot{x} = x\,(1-x) + \frac{f(q-x)}{q+x}\,y \\ \dot{y} = x - y \end{cases} \tag{1.3}$$

By using $\varepsilon = 4 \times 10^{-4}$, $f = 2/3$ and $q = 8 \times 10^{-2}$, system (1.3) produces a limit cycle for different sets of initial conditions.

As it is observed from the aforementioned systems, their structure was considered such that the existence of oscillations was "almost obvious." This means that the oscillation was excited from an unstable equilibrium, which nowadays is known as "self-excited" oscillation. From a computational point of view, this allows one to use a standard computational procedure, in which after a transient process, a trajectory, starting from a point of unstable manifold in a neighborhood of equilibrium, reaches a state of oscillation.

Therefore, by considering a dynamical system

$$\dot{X} = F\,(X, p) \tag{1.4}$$

where $X \in R^n$, $t \in R$, and $p \in R^k$ are the vectors of system parameters. From a computational point of view, it is natural to introduce the following class of attractors:

Definition 1.1 [66, 78] An attractor is called a self-excited attractor if its basin of attraction intersects with any open neighborhood of an unstable fixed point.

Furthermore, in the middle of twentieth century, except for self-excited periodic oscillations in applied systems, chaotic oscillations were found numerically to be also excited from an unstable equilibrium and can be computed by the standard computational procedure. The most representative members of this class of systems are presented in detail.

In 1963, Edward Lorenz developed a simplified mathematical model for atmospheric convection [91]. This model is a system of three ordinary differential equations now known as the Lorenz equations:

$$\begin{cases} \dot{x} = \sigma \left(y - x \right) \\ \dot{y} = x \left(\rho - z \right) - y \\ \dot{z} = xy - \beta z \end{cases} \tag{1.5}$$

The Lorenz system was the first well-known example of a visualization of chaotic attractor in a dynamical system corresponding to the excitation of chaotic attractor from unstable equilibria. For classical parameters ($\sigma = 10$, $\beta = 8/3$, $\rho = 28$), the Lorenz attractor is self-excited with respect to all three equilibria and could have been found using the standard computational procedure (see Fig. 1.3). The Lorenz equations also arise in simplified models for lasers [37], dynamos [61], thermosyphons [30], brushless DC motors [39], electric circuits [18], chemical reactions [115], and forward osmosis [143].

Some years later, Rabinovich proposed a system [110, 119] which describes the interaction of plasma waves and was based on the generalized Lorenz system

$$\begin{cases} \dot{x} = -\sigma \left(x - y \right) - ayz \\ \dot{y} = rx - y - xz \\ \dot{z} = xy - bz \end{cases} \tag{1.6}$$

Fig. 1.3 Phase portrait in the $x - y$ plane for the Lorenz system, with $\sigma = 10$, $\beta = 8/3$, $\rho = 28$, for initial conditions $(x(0), y(0), z(0)) = (1, 1, 1)$

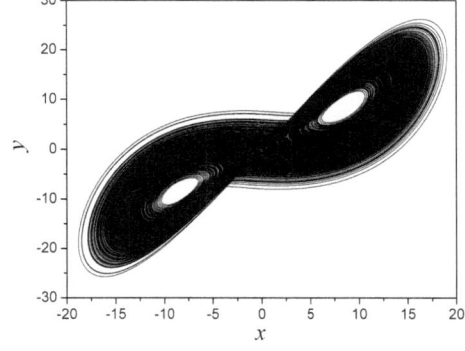

Fig. 1.4 Phase portrait in
the $x - y$ plane for the
Rabinovich system, with
$\sigma = 4, a = 0.35, b = 1$, and
$r = 11.39$, for initial
conditions
$(x(0), y(0), z(0)) = (1, 1, 1)$

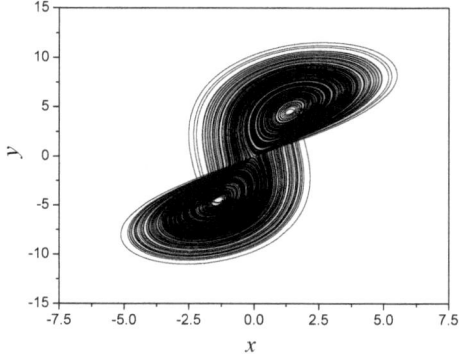

with $a < 0$. For positive $a > 0$, it corresponds to the Glukhovsky–Dolghansky system, which describes convective fluid motion and was considered in 1980 [27]. Also, it describes a rigid body rotation in a resisting medium and the forced motion of a gyrostat [119]. By using $\sigma = 4$, $a = 0.35$, $b = 1$ and $r = 11.39$, system (1.6) produces a chaotic attractor as presented in Fig. 1.4.

Another interesting system with chaotic behavior is the system for the classical Chua's oscillator circuit, which is described as [28]:

$$\begin{cases} \dot{x} = a\,(y - x) - af(x) \\ \dot{y} = x - y + z \\ \dot{z} = -(\beta y + \gamma z) \end{cases} \tag{1.7}$$

where the dimensionless form of the nonlinear function $f(x)$ of the Chua's diode is given by the following equation:

$$f(x) = m_c x + 0.5\,(m_a - m_b)\,(|x + 1| - |x - 1|) + 0.5\,(m_b - m_c)\,(|x + E_2/E_1| - |x - E_2/E_1|) \tag{1.8}$$

The Chua's oscillator circuit is structurally the simplest and dynamically the most complex member of a larger class of nonlinear circuits, known as Chua's Circuit Family. In Fig. 1.5, an experimental chaotic phase portrait, produced by the Chua's oscillator circuit, is illustrated.

This series of examples of self-excited attractors can be extended to the case of other well-known dynamical systems. Nowadays, thousands of publications have been devoted to the computation and analysis of systems of this class.

1.2 Hidden Oscillations

A further study showed that the self-excited periodic and chaotic oscillations did not give exhaustive information about the possible types of oscillations. In the middle of twentieth century, the examples of periodic and chaotic oscillations of another

Fig. 1.5 Experimental
chaotic phase portrait in the
$x - y$ plane for the Chua's
oscillator circuit

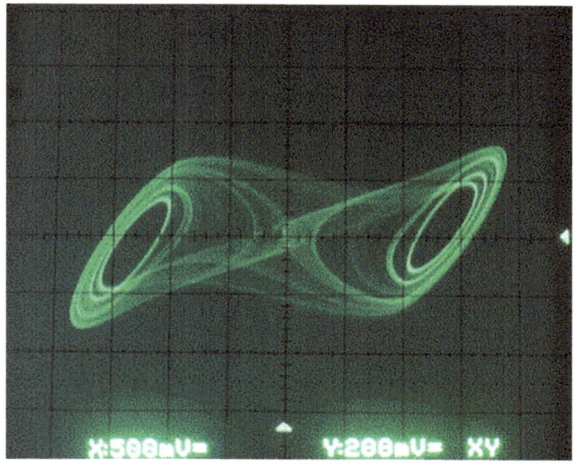

type were found, later called [80] the "hidden oscillations" and "hidden attractors," of which the basin of attraction does not intersect with small neighborhoods of equilibria. So, this class of attractors should be introduced according to the following definition.

Definition 1.2 An attractor is called a hidden attractor if its basin of attraction does not intersect with small neighborhoods of equilibria.

Nowadays, the problem of numerical localization, computation, and analytical investigation of hidden attractors is much more challenging. This happens, since in this case there is no possibility to use information about equilibria for organization of similar transient processes in the standard computational procedure. Thus, the hidden attractors cannot be computed by using this standard procedure. Furthermore, in this case it is unlikely that the integration of trajectories into random initial data furnishes hidden attractor localization since a basin of attraction can be very small and the dimension of hidden attractor itself can be much lesser than the dimension of the considered system.

The first who posed the problem of investigating hidden oscillations is David Hilbert. In 1900, he introduced the problem of the investigation of the number and possible dispositions of limit cycles in two-dimensional polynomial systems in relation to the degree of the considered polynomials. This is the second part of 16th Hilbert Problem [40], which Smale reformulated later in the following way [129]:
"Is there a bound $K = H(n)$ on the number of limit cycles of the form $K < n^q$ for the polynomial system (1.9), where n is the maximum of the degree of polynomials P_n and Q_n and q is a universal constant."

$$\begin{cases} \dot{x} = P_n(x, y) \\ \dot{y} = Q_n(x, y) \end{cases} \tag{1.9}$$

For more than a century, in attempting to solve this problem, numerous analytical results were obtained. But the problem is still far from being resolved even for a simple class of quadratic systems.

The first non-trivial results have been obtained by the creation of effective methods for the construction of systems with limit cycles, which were initiated by Bautin [5–7]. In his works, for the construction of nested limit cycles, an effective analytical method was proposed based on determining the sequential symbolic expressions of Lyapunov values (called also focus values, Lyapunov quantities, Lyapunov coefficients, Poincaré–Lyapunov constants). The sequential computation of Lyapunov values, used by Bautin, allowed him to discriminate first a class of quadratic systems, in which three nested limit cycles can be found in the neighborhood of degenerate focus by small perturbations of system coefficients [5] (such cycles are naturally called small or small-amplitude limit cycles). Next, Petrovskii and Landis [104] asserted that quadratic system can have less than or equal to three limit cycles. But later they reported a gap in the proof [105]. Quadratic systems were found with four limit cycles [16, 127] (three nested small limit cycles, obtained by Bautins technique, and one large (or normal amplitude) limit cycle, surrounding another focus equilibrium).

So, up to now, the best result of possible estimation for number $H(2)$ of limit cycles in quadratic system is $H(2) \geq 4$ and it is finite (for cubic systems $H(3) \geq 13$) [85]; in the work of Ref. [38], a lower estimate is given for the Hilbert number $H(n)$: It grows at least as rapidly as $(2 \ln 2)^{-1}(n + 2)^2 \ln (n + 2)$ for all large n.

The Definition 1.2 of hidden attractors refers to the autonomous system, but it can also be generalized to the non-autonomous dynamical systems [78]. Let us consider Poincaré map P, where points are taken by the period of external excitation $2\pi/\omega$, i.e., $P = \{(x(t), \dot{x}(t)) : t = 2n\pi/\omega, \, n = 0, 1, \ldots\}$.

Definition 1.3 The attractor of a non-autonomous dynamical system is hidden if its basin of attraction does not touch the neighborhood of the equilibrium on such defined Poincaré section.

The appearance of modern computers permits one to use numerical simulation of complicated nonlinear dynamical systems and to obtain new information on a structure of their trajectories. However, the possibilities of "simple" approach, based on the construction of trajectories by numerical integration of the considered differential equations, turned out to be highly limited.

Furthermore, during the last few years the study of dynamical systems with hidden attractors has attracted the research interest due to a number of possible applications of such systems in various fields.

1.3 Localization of Hidden Attractors

One of the effective methods for the numerical localization of hidden attractors is based on a homotopy and numerical continuation [20]. We construct a sequence of similar systems such that the initial data for numerically computing the oscillating solution (starting oscillation) can be obtained analytically for the first (starting) system. For example, it is often possible to consider a starting system with a self-excited starting oscillation. Then, we numerically track the transformation of the starting oscillation while passing between the systems.

In a scenario of transition to chaos in the dynamical system, there is a typical parameter $\lambda \in [a_1, a_2]$, the variation of which gives the scenario. We can also artificially introduce parameter λ, let it vary in the interval $[a_1, a_2]$ (where $\lambda = a_2$ corresponds to the initial system) and choose parameter a_1 such that we can analytically or computationally find a certain non-trivial attractor when $\lambda = a_1$ (this attractor has often a simple form, e.g., periodic). That is, instead of analyzing the scenario of a transition into chaos, we can synthesize it. Further, we consider the sequence λ_j, $\lambda_1 = a_1$, $\lambda_m = a_2$, $\lambda_j \in [a_1, a_2]$ such that the distance between λ_j and $\lambda_j + 1$ is sufficiently small. Then, we numerically investigate the changes in the shape of the attractor obtained for $\lambda_1 = a_1$. If the change in λ (from λ_j to λ_{j+1}) does not cause a loss of the stability bifurcation of the considered attractor, then the attractor for $\lambda_m = a_2$ (at the end of procedure) is localized.

1.4 Control and Synchronization

Over the last few decades, the control and synchronization of oscillating motions in dynamical systems have been the topic of intense research from both theoretical and experimental points of view [100, 117, 118, 124, 128, 141]. In many practical situations, fixed point solution is desirable, for example in laser applications [26, 55, 64, 65] where a constant output is needed and fluctuations should be avoided. There are also other situations where oscillations need to be maintained, for instance in brain functioning [44, 123]. In some systems, the regular oscillation is necessary rather than the irregular one, e.g., beating of heart [17]. Similarly, in telecommunication chaotic signal is used as a carrier signal rather than a periodic one [53]. These different requirements for specific types of motion suggest that appropriate control strategies are necessary.

The control of chaotic motion to periodic one is the most difficult due to an extreme sensitivity to initial conditions. However, various approaches have been established to control such motion. Foremost is the OGY method named after Ott, Grebogi, and Yorke [101], where unstable periodic orbits are stabilized using a linear feedback method. This method generated widespread interest, and its applications can be found in almost all branches of science [29, 124]. Another technique for chaos control is the time-delay feedback proposed by Pyragas [117], where unstable periodic orbits of a chaotic system are stabilized by the use of a specially designed external oscillator or by delayed self-controlling feedback without using any external force. There are other equally important methods, such as active control, passive control, sampled-data feedback control, sliding mode control, backstepping control, Takagi–Sugeno, fuzzy control, or adaptive control [43, 146, 147, 165–168].

Over the last two decades, controlling the oscillating motion to the unstable fixed points has also been a topic of intense research from both theoretical and experimental points of view [100]. Note that all the aforementioned schemes for controlling dynamical motion have been developed and used on the systems having self-excited oscillation. However, during the last few years a number of works on controlling chaotic systems with hidden attractors have been reported in the literature [56, 126, 145, 158].

Furthermore, during the last three decades the phenomenon of synchronization between coupled chaotic systems has attracted the interest of the scientific community because it is a rich and multi-disciplinary phenomenon with broad range applications, such as in secure communications [45] and cryptography [33, 149], in broadband communications systems [19], and in a variety of complex physical, chemical, and biological systems [41, 90, 98, 111, 137, 140, 151]. In general, synchronization of chaos is a process, where two or more chaotic systems adjust a given property of their motion to a common behavior, such as equal trajectories or phase locking, due to coupling or forcing. Because of the exponential divergence of the nearby trajectories of a chaotic system, having two chaotic systems being synchronized might be a surprise. However, today the synchronization of coupled chaotic oscillators is a phenomenon that is well established experimentally and reasonably well understood theoretically.

The history of chaotic synchronization's theory began with the study of the interaction between coupled chaotic systems in the 1980s and early 1990s by Fujisaka and Yamada [25], Pikovsky [112], Pecora and Carroll [103]. Since then, a wide range of research activity based on the synchronization of nonlinear systems has risen, and a variety of synchronization's forms depending on the nature of the interacting systems and of the coupling schemes has been presented. Complete or full chaotic synchronization [49, 68, 69, 71, 72, 94, 112, 161, 163], phase synchronization [21, 102], lag synchronization [121, 138], generalized synchronization [122], anti-synchronization [54, 89] and anti-phase synchronization [4, 11, 15, 70, 142, 170], projective synchronization [92], anticipating [150], and inverse lag synchronization [88] are the most interesting types of synchronization that have been investigated numerically and experimentally by many research groups.

One of the simplest synchronization schemes, which have been reported, is the synchronization scheme via diffusion coupling [35, 98, 164]. The diffusive coupling is the most common type of coupling appearing in real systems. Furthermore, from the control point of view, the synchronization issue deals with the synchronization of a pair of systems named the master and the slave systems [12, 36]. So, the aim of this approach is to design the control laws which guarantee that the output of the slave systems tracks the output of the master system [48]. For this reason, various methodologies have been investigated for the synchronization of systems such as active control, passive control, sampled-data feedback control, sliding mode control, backstepping control, Takagi–Sugeno, fuzzy control, or adaptive control [169]. During the last few years, a number of works on the synchronization of chaotic systems with hidden attractors have been reported in the literature [108, 109, 125, 144].

1.5 Hidden Oscillations in Applied Models

In this section, a brief synopsis of few examples of hidden oscillations in applied models is considered.

1.5.1 Phase-Locked Loop Circuits

The phase-locked loop (PLL) systems were invented in the 1930s–1940s [8] and were widely used in radio and television (demodulation and recovery, synchronization and frequency synthesis). Nowadays, PLL can be produced in the form of single integrated circuit, and the various modifications of PLL are used in a variety of modern electronic applications (radio, telecommunications, computers, and others). Various methods for the analysis of PLLs have been well developed by engineers and are considered in many publications [162]. However, the problems of the construction of adequate nonlinear models as well as the nonlinear analysis of such models, which are still far from being resolved, turn out to be difficult and require to use special methods of the qualitative theory of differential, difference, integral, and integro-differential equations [63, 67, 79, 81–83, 93, 107, 136, 148].

In the middle of the last century, Kapranov studied [52] qualitative behavior of PLL systems. In these investigations, Kapranov assumed that in PLL system

$$\begin{cases} \dot{z} = -\alpha z - (1 - a\alpha)\left(\varphi\left(\sigma\right) - \gamma\right) \\ \dot{\sigma} = z - a\left(\varphi\left(\sigma\right) - \gamma\right) \end{cases} \tag{1.10}$$

with $a, \alpha, \gamma \geq 0$, where $\varphi(\alpha)$ is a 2π–periodic characteristic of the phase detector and with the filter of type $W\left(p\right) = \frac{ap+\beta}{p+a}$ there were self-excited oscillations only.

However, in 1961, Gubar' [34] revealed a gap in Kapranov's work and showed analytically the possibility of the existence of hidden oscillations in two-dimensional system of PLL. Thus, from a computational point of view, the system considered was globally stable (all the trajectories tend to equilibria), but, in fact, there was a bounded domain of attraction only.

1.5.2 Automatic Control Systems

In the 1950–1960s, the investigations of widely known Markus–Yamabe [95], Aizerman [1], and Kalman [50] conjectures on absolute stability have led to the finding
of hidden oscillations in automatic control systems with a unique stable stationary
point and with a nonlinearity, which belongs to the sector of linear stability [10, 13,
24, 62, 75, 113].

At the end of the last century, the difficulties of numerical analysis of hidden
oscillations arose in the simulation of aircraft's control systems (anti-windup scheme)
and caused the crash of aircraft YF–22 Boeing in April 1992 [74]. Therefore, a crash
like this has shown that it is necessary to develop special effective methods, since the
stability in simulations does not imply the stability of the physical control system
and also stronger theoretical understanding is required.

1.5.3 Chua's Circuit Oscillator

In 2010, the discovery, for the first time, of chaotic hidden attractor in a generalized
Chua's circuit [66, 84] and later discovery of chaotic hidden attractor in classical
Chua's circuit [80] have been reported. It should be remarked that for the last thirty
years, several thousand publications, in which a few hundreds of attractors were discussed, have been devoted to Chua's circuit and its various modifications. However,
up to now these Chua's attractors were self-excited.

1.5.4 Electromechanical Systems

Hidden attractors appear naturally in systems without equilibria, describing various
electromechanical models with rotation and electrical circuits with cylindrical phase
space. One of the first such examples has been described by Arnold Sommerfeld in
1902 [130]. He has studied the oscillations caused by a motor driving an unbalanced
weight and discovered the resonance capture (Sommerfeld effect). The Sommerfeld
effect of the capture phenomenon represents the failure of a rotating mechanical
system to be spun up by a torque-limited rotor to a desired rotational velocity due to
its resonant interaction with another part of the system [22, 23].

Furthermore, in Refs. [14, 96] a double-mass mathematical model of the drilling
system, which is based on an experimental setup, is studied. It consists of upper and
lower disks connected with each other by a steel string. The upper disk is actuated
by a DC motor, and there is also a brake device which is used for the modeling of
the friction force which acts on the lower disc.

In the aforementioned dynamical model of the drilling system there are two natural transition processes: The first is to start the motor and then to begin drilling (i.e., to add friction to the model), and the second is to apply friction to the model and then to start the motor (also it corresponds to the sudden changes in the friction). The first transition process leads to the normal operation (i.e., the corresponding trajectory is attracted to the stable stationary point), while the second transition process may lead to a hidden oscillation. Similar behavior and hidden periodic attractors are also found in the drilling systems driven by induction motors with a wound rotor or salient pole synchronous motors [58–60, 76].

1.6 Families of Systems with Hidden Attractors

In this section, the main families of systems with hidden attractors will be presented.

1.6.1 Systems Without Equilibrium

The works of Nosé [99] and Hoover [42] in 1984–1985 have led the study of the following dynamical system without equilibria

$$\begin{cases} \dot{x} = y \\ \dot{y} = -x - yz \\ \dot{z} = a\left(y^2 - 1\right) \end{cases} \tag{1.11}$$

and its various modifications, where hidden chaotic oscillations can be found [116, 131–133, 152].

Systematic search routine was developed by Jafari et al. to determine simple quadratic flows with no equilibria [46, 154]. Wang and Chen found a new system without equilibrium while studying a chaotic system with any number of equilibria [154]. Wei discovered dynamical properties of a no-equilibrium chaotic system by applying a constant to the Sprott D system [157]. Multiple attractors in a three-dimensional system with no-equilibrium point were reported in [171]. Akgul et al. designed a random number generator with a 3D chaotic system without equilibrium point [2]. In addition, 4D no-equilibrium systems with hyperchaos were presented in [155, 156, 159]. These examples motivate further the construction and study of various other chaotic systems without equilibria [51, 77, 80].

1.6.2   Systems with Stable Equilibrium

The example of such unusual chaotic flow with only one stable equilibrium has been designed by Wang and Chen [153]. They have considered the following system:

$$\begin{cases} \dot{x} = yz + 0.006 \\ \dot{y} = x^2 - y \\ \dot{z} = 1 - 4x \end{cases} \tag{1.12}$$

A general model of Wang–Chen system was studied in [160]. By using a search routine, quadratic chaotic flows with one stable equilibrium were discovered by Molaie et al. [97]. Kingni et al. introduced integer-order and fractional-order forms of a new three-dimensional chaotic system with one stable equilibrium [56]. Moreover, a cost function based on Gaussian mixture model was developed to estimate the parameters of a chaotic circuit with stable equilibrium [73].

1.6.3   Systems with an Infinite Number of Equilibria

There are two main research directions relating to chaotic systems with an infinite number of equilibrium points.

In the first one, after proposing a chaotic system with any number of equilibria by Wang and Chen in [153], Jafari and Sprott in [47] have introduced simple chaotic systems with a line of equilibria. They have been inspired by the structure of the conservative Sprott case A system [131] and have considered a general parametric form of it with quadratic nonlinearities. With exhaustive computer search, nine simple cases have been found. The system given by the following equations:

$$\begin{cases} \dot{x} = y \\ \dot{y} = -x + yz \\ \dot{z} = -x - 15xy - xz \end{cases} \tag{1.13}$$

is an especially simple example with only six terms. System (1.13) has a line of stable equilibria at $(0, 0, z^*)$, where $z^* \in R$, with no other equilibria. So, the z-axis is the line equilibrium of this system.

Later, five new chaotic flows with a line equilibrium and especially a complicated one with two infinite parallel lines of equilibrium points were discovered by Li and Sprott [86]. By using signum functions and absolute-value functions, chaotic systems with a line or two perpendicular lines of equilibrium points were introduced in [87].

Furthermore, after the discovery of a new class of chaotic systems with circular equilibrium [31], the authors concentrated on chaotic systems with curve equilibrium. Kingni et al. presented a 3D chaotic autonomous system with a circular equilibrium and its fractional-order form [57]. Gotthans et al. reported another chaotic system with circular equilibrium in [32]. Moreover, a 3D system with a square equilibrium, which was constructed by modifying the system with circular equilibrium [32], is also proposed. In addition, a system exhibiting chaotic attractor with ellipse equilibrium, chaotic attractor with square-shaped equilibrium, and chaotic attractor with rectangle-shaped equilibrium was represented in [106].

References

1. Aizerman, M.A.: On a problem concerning the stability in the large of dynamical systems. Uspekhi Mat. Nauk **4**, 187–188 (1949)
2. Akgul, A., Calgan, H., Koyuncu, I., Pehlivan, I., Istanbullu, A.: Chaos-based engineering applications with a 3D chaotic system without equilibrium points. Nonlinear Dyn. **84**, 481–495 (2016)
3. Andronov, A.A., Khaikin, S.E.: Theory of Oscillators. Pergamon, Oxford (1966)
4. Astakhov, V., Shabunin, A., Anishchenko, V.: Antiphase synchronization in symmetrically coupled self-oscillators. Int. J. Bifurc. Chaos **10**, 849–857 (2000)
5. Bautin, N.N.: On the number of limit cycles appearing on varying the coefficients from a focus or centre type of equilibrium state. Mat. Sb. (N.S.) **30**, 181–196 (1939)
6. Bautin, N.N.: On the number of limit cycles generated on varying the coefficients from a focus or centre type equilibrium state. Doklady Akademii Nauk SSSR **24**, 668–671 (1939)
7. Bautin, N.N.: The Behaviour of Dynamical Systems Close to the Boundaries of a Stability Domain. Gostekhizdat, Leningrad, Moscow (1949)
8. Bellescize, H.: La réception synchrone. Londe Electrique **24**, 230–340 (1932)
9. Belousov, B.P.: A periodic reaction and its mechanism. Collection of Short Papers on Radiation Medicine for 1958. Medknow Publications, Moscow (1959)
10. Bernat, J., Llibre, J.: Counterexample to Kalman and Markus-Yamabe conjectures in dimension larger than 3. Dyn. Contin. Discr. Impul. Syst. **2**, 337–379 (1996)
11. Blazejczuk-Okolewska, B., Brindley, J., Czolczynski, K., Kapitaniak, T.: Antiphase synchronization of chaos by noncontinuous coupling: two impacting oscillators. Chaos Solitons Fract. **12**, 1823–1826 (2001)
12. Boccaletti, S., Kurths, J., Osipov, G., Valladares, D., Zhou, C.: The synchronization of chaotic system. Phys. Rep. **366**, 1–101 (2002)
13. Bragin, V.O., Vagaitsev, V.I., Kuznetsov, N.V., Leonov, G.A.: Algorithms for finding hidden oscillations in nonlinear systems. the Aizerman and Kalman conjectures and Chua's circuits. J. Comput. Syst. Sci. Int. **50**, 511–543 (2011)
14. de Bruin, J., Doris, A., van de Wouw, N., Heemels, W., Nijmeijer, H.: Control of mechanical motion systems with non-collocation of actuation and friction: a Popov criterion approach for input-to-state stability and set-valued nonlinearities. Automatica **45**, 405–415 (2009)
15. Cao, L.Y., Lai, Y.C.: Antiphase synchronism in chaotic system. Phys. Rev. **58**, 382–386 (1998)
16. Chen, L.S., Wang, M.S.: The relative position and number of limit cycles of the quadratic differential systems. Acta Math. Sin. **22**, 751–758 (1979)
17. Christini, D.J., Stein, K.M., Markowitz, S.M., Mittal, S., Slotwiner, D.J., Scheiner, M.A., Iwai, S., Lerma, B.B.: Nonlinear-dynamical arrhythmia control in humans. Proc. Natl. Acad. Sci. USA **98**, 5827–5832 (2001)
18. Cuomo, K.M., Oppenheim, A.V.: Circuit implementation of synchronized chaos with applications to communications. Phys. Rev. Lett. **71**, 65–68 (1993)
19. Dimitriev, A.S., Kletsovi, A.V., Laktyushkin, A.M., Panas, A.I., Starkov, S.O.: Ultrawideband wireless communications based on dynamic chaos. J. Commun. Technol. Electron. **51**, 1126–1140 (2006)
20. Dudkowski, D., Jafari, S., Kapitaniak, T., Kuznetsov, N., Leonov, G., Prasad, A.: Hidden attractors in dynamical systems. Phys. Rep. **637**, 1–50 (2016)

21. Dykman, G.I., Landa, P.S., Neymark, Y.I.: Synchronizing the chaotic oscillations by external force. Chaos Solitons Fract. **1**, 339–353 (1991)
22. Eckert, M.: Arnold Sommerfeld: Science, Life and Turbulent Times 1868–1951. Springer, New York (2013)
23. Evan-Iwanowski, R.: Resonance Oscillations in Mechanical Systems. Elsevier (1976)
24. Fitts, R.E.: Two counterexamples to Aizerman's conjecture. IEEE Trans. Autom. Control **11**, 553–556 (1966)
25. Fujisaka, H., Yamada, T.: Stability theory of synchronized motion in coupled-oscillator systems. Prog. Theory Phys. **69**, 32–47 (1982)
26. Gangwar, V.P., Prasad, A., Ghosh, R.: Optical phase dynamics in mutually coupled diode laser systems exhibiting power synchronization. J. Phys. B **44**, 235,403 (2011)
27. Glukhovskii, A.B., Dolzhanskii, F.V.: Three-component geostrophic model of convection in a rotating fluid. Acad. Sci. USSR Izv. Atmos. Ocean. Phys. **16**, 311–318 (1980)
28. Glukhovskii, A.B., Dolzhanskii, F.V.: A chaotic attractor from Chua's circuit. IEEE Trans. Circuits Syst. **31**, 1055–1058 (1990)
29. Gonzalez-Miranda, J.M.: Synchronization and Control of Chaos: An Introduction for Scientists and Engineers. World Scientific, Singapore (2004)
30. Gorman, M., Widmann, P.J., Robbins, K.: Nonlinear dynamics of a convection loop: a quantitative comparison of experiment with theory. Physica D **19**, 255–267 (1986)
31. Gotthans, T., Petržela, J.: New class of chaotic systems with circular equilibrium. Nonlinear Dyn. **73**, 429–436 (2015)
32. Gotthans, T., Sportt, J.C., Petržela, J.: Simple chaotic flow with circle and square equilibrium. Int. J. Bifurc. Chaos **26**(1650), 137–138 (2016)
33. Grassi, G., Mascolo, S.: Synchronization of high-order oscillators by observer design with application to hyperchaos-based cryptography. Int. J. Circuit Theory Appl. **27**, 543–553 (1999)
34. Gubar, N.A.: Investigation of a piecewise linear dynamical system with three parameters. J. Appl. Math. Mech. **25**, 1011–1023 (1961)
35. Guemez, J., Matias, M.A.: Modified method for synchronizing and cascading chaotic system. Phys. Rev. E **61**, R2145–R2148 (1995)
36. Guemez, J., Matias, M.A.: Modified method for synchronizing and cascading chaotic system. Phys. Rev. E **52**, R2145–R2148 (1995)
37. Haken, H.: Analogy between higher instabilities in fluids and lasers. Phys. Lett. A **53**, 77–78 (1975)
38. Han, M., Li, J.: Lowerbounds for the Hilbert number of polynomial systems. J. Differ. Equ. **252**, 3278–3304 (2012)
39. Hemati, N.: Strange attractors in brushless DC motors. IEEE Trans. Circuit Syst. I: Fund. Theory Appl. **41**, 40–45 (1994)
40. Hilbert, D.: Mathematical problems. Bull. Am. Math. Soc. **8**, 437–479 (1902)
41. Holstein-Rathlou, N.H., Yip, K.P., Sosnovtseva, O.V., Mosekilde, E.: Synchronization phenomena in nephron-nephron interaction. Chaos **11**, 417–426 (2001)
42. Hoover, W.: Canonical dynamics: equilibrium phase-space distributions. Phys. Rev. A **31**, 1695 (1985)
43. Hua, C., Guan, X.: Adaptive control for chaotic systems. Chaos Solitons Fract. **22**, 55–60 (2004)
44. Iasemidis, L.D.: Epileptic seizure prediction and control. IEEE Trans. Biomed. Eng. **50**, 549–558 (2003)
45. Jafari, S., Haeri, M., Tavazoei, M.S.: Experimental study of a chaos-based communication system in the presence of unknown transmission delay. Int. J. Circuit Theory Appl. **38**, 1013–1025 (2010)
46. Jafari, S., Sprott, J., Golpayegani, S.M.R.H.: Elementary quadratic chaotic flows with no equilibria. Phys. Lett. A **377**, 699–702 (2013)
47. Jafari, S., Sprott, J.C.: Simple chaotic flows with a line equilibrium. Chaos Solitons Fract. **57**, 79–84 (2013)

48. Jovic, B.: Synchronization Techniques for Chaotic Communication Systems. Springer, Germany (2011)
49. Kadji, H.G.E., Orou, J.B.C., Woafo, P.: Synchronization dynamics in a ring of four mutually coupled biological systems. Commun. Nonlinear Sci. Numer. Simul. **13**, 1361–1372 (2008)
50. Kalman, R.E.: Physical and mathematical mechanisms of instability in nonlinear automatic control systems. Trans. ASME **79**, 553–566 (1957)
51. Kapitaniak, T., Leonov, G.A.: Multistability: uncovering hidden attractors. Eur. Phys. J. Spec. Top. **224**, 1405–1408 (2015)
52. Kapranov, M.: Locking band for phase-locked loop. Radiofizika **2**, 37–52 (1956)
53. Kennedy, M., Rovatti, R., Setti, G.: Chaotic Electronics in Telecommunications. CRC Press, USA (2000)
54. Kim, C.M., Rim, S., Kye, W.H., Rye, J.W., Park, Y.J.: Anti-synchronization of chaotic oscillators. Phys. Lett. A **320**, 39–46 (2003)
55. Kim, M.Y., Roy, R., Aron, J.L., Carr, T.W., Schwartz, I.B.: Scaling behavior of laser population dynamics with time–delayed coupling: theory and experiment. Phys. Rev. Lett. **94**, 088,101 (2005)
56. Kingni, S.T., Jafari, S., Simo, H., Woafo, P.: Three-dimensional chaotic autonomous system with only one stable equilibrium: analysis, circuit design, parameter estimation, control, synchronization and its fractional-order form. Eur. Phys. J. Plus **129**, 76 (2014)
57. Kingni, S.T., Pham, V.T., Jafari, S., Kol, G.R., Woafo, P.: Three-dimensional chaotic autonomous system with a circular equilibrium: analysis, circuit implementation and its fractional-order form. Circuits Syst. Signal Process **35**(19), 331–1948 (2016)
58. Kiseleva, M., Kondratyeva, N., Kuznetsov, N., Leonov, G.: Hidden oscillations in drilling systems with salient pole synchronous motor. IFAC Proc. **48**, 700–705 (2015)
59. Kiseleva, M., Kondratyeva, N., Kuznetsov, N., Leonov, G., Solovyeva, E.: Hidden periodic oscillations in drilling system driven by induction motor. IFAC Proc. **19**, 5872–5877 (2014)
60. Kiseleva, M., Kuznetsov, N.V., Leonov, G.A., Neittaanmaki, P.: Hidden oscillations in drilling system actuated by induction motor. IFAC Proc. **5**, 86–89 (2013)
61. Knobloch, E.: Chaos in the segmented disc dynamo. Phys. Lett. A **82**, 439–440 (1981)
62. Krasovsky, N.N.: Theorems on the stability of motions determined by a system of two equations. Prikl. Mat. Mekh. **16**, 547–554 (1952)
63. Kudrewicz, J., Wasowicz, S.: Equations of Phase Locked Loop Dynamics on Circle. Torus and Cylinder. World Scientific, Singapore (2007)
64. Kumar, P., Prasad, A., Ghosh, R.: Stable phase–locking of an external cavity diode laser. J. Phys. B **41**, 135,402 (2008)
65. Kumar, P., Prasad, A., Ghosh, R.: Strange bifurcation and phase–locked dynamics in mutually coupled diode laser systems. J. Phys. B **42**, 145,401 (2009)
66. Kuznetsov, N.V., Leonov, G.A., Vagaitsev, V.: Analytical-numerical method for attractor localization of generalized Chua's system. IFAC Proc. **4**, 29–33 (2010)
67. Kuznetsov, N.V., Leonov, G.A., Yuldashev, M.V., Yuldashev, R.V.: Analytical methods for computation of phase–detector characteristics and PLL design. In: Proceedings of International Symposium on Signals, Circuits and Systems (ISSCS'2011), Iasi, Romania, pp. 7–10 (2011)
68. Kyprianidis, I.M., Stouboulos, I.N.: Chaotic synchronization of three coupled oscillators with ring connection. Chaos Solitons Fract. **17**, 327–336 (2003)
69. Kyprianidis, I.M., Stouboulos, I.N.: Synchronization of two resistively coupled nonautonomous and hyperchaotic oscillators. Chaos Solitons Fract. **17**, 317–325 (2003)
70. Kyprianidis, I.M., Bogiatzi, A.N., Papadopoulou, M., Stouboulos, I.N., Bogiatzis, G.N., Bountis, T.: Synchronizing chaotic attractors of chuas canonical circuit. The case of uncertainty in chaos synchronization. Int. J. Bifurc Chaos **16**, 1961–1976 (2006)
71. Kyprianidis, I.M., Volos, C.K., Stouboulos, I.N.: Experimental synchronization of two resistively coupled Duffing-type circuits. Nonlinear Phenom. Complex Syst. **11**, 187–192 (2008)
72. Kyprianidis, I.M., Volos, C.K., Stouboulos, I.N., Hadjidemetriou, J.: Dynamics of two resistively coupled Duffing-type electrical oscillators. Int. J. Bifurc. Chaos **16**, 1765–1775 (2006)

73. Lao, S.K., Shekofteh, Y., Jafari, S., Sprott, J.C.: Cost function based on Gaussian mixture model for parameter estimation of a chaotic circuit with a hidden attractor. Int. J. Bifurc. Chaos **24**, 1450,010 (2014)

74. Lauvdal, T., Murray, R., Fossen, T.: Stabilization of integrator chains in the presence of magnitude and rate saturations: a gain scheduling approach. In: Proceedings of IEEE Control and Decision Conference (CDC'97), San Diego, USA, pp. 4404–4005 (1997)

75. Leonov, G.A., Kuznetsov, N.V.: Numerical Methods for Differential Equations, Optimization, and Technological Problems. *Computational Methods in Applied Sciences*, vol. 27, chap. Analytical–numerical methods for hidden attractors' localization: the 16th Hilbert problem, Aizerman and Kalman conjectures, and Chua circuits, pp. 41–64. Springer International Publishing (2013)

76. Leonov, G.A., Kuznetsov, N.V., Kiseleva, M.A., Solovyeva, E.P., Zaretskiy, A.M.: Hidden oscillations in mathematical model of drilling system actuated by induction motor with a wound rotor. Nonlinear Dyn. **77**, 277–288 (2014)

77. Leonov, G.A., Kuznetsov, N.V., Kuznetsova, O.A., Seldedzhi, S.M., Vagaitsev, V.I.: Hidden oscillations in dynamical systems. Trans. Syst. Control **6**, 54–67 (2011)

78. Leonov, G.A., Kuznetsov, N.V., Mokaev, T.N.: Homoclinic orbits, and self-excited and hidden attractors in a Lorenz-like system describing convective fluid motion–homoclinic orbits, and self-excited and hidden attractors. Eur. Phys. J. Spec. Top. **224**, 1421–1458 (2015)

79. Leonov, G.A., Kuznetsov, N.V., Seledzhi, S.M.: Automation Control—Theory and Practice, chap. Nonlinear analysis and design of phase-locked loops, pp. 89–114. Robotics and Automation. InTech (2009)

80. Leonov, G.A., Kuznetsov, N.V., Vagaitsev, V.I.: Localization of hidden Chua's attractors. Phys. Lett. A **375**, 2230–2233 (2011)

81. Leonov, G.A., Kuznetsov, N.V., Yuldahsev, M.V., Yuldashev, R.V.: Computation of phase detector characteristics in synchronization systems. Dokl. Math. **84**, 586–590 (2011)

82. Leonov, G.A., Kuznetsov, N.V., Yuldahsev, M.V., Yuldashev, R.V.: Analytical method for computation of phase-detector characteristic. IEEE Trans. Circuits Syst.-II: Express. Briefs **59**, 633–647 (2012)

83. Leonov, G.A., Ponomareko, D.V., Smirnova, V.B.: Frequency-Domain Methods for Nonlinear Analysis: Theory and Applications. World Scientific, Singapore (1996)

84. Leonov, G.A., Vagaitsev, V.I., Kuznetsov, N.V.: Algorithm for localizing Chua attractors based on the harmonic linearization method. Dokl. Math. **82**, 693–696 (2010)

85. Li, C., Liu, C., Yanga, J.: A cubic system with thirteen limit cycles. J. Differ. Equ. **246**, 3609–3619 (2009)

86. Li, C., Sprott, J.C.: Chaotic flows with a single nonquadratic term. Phys. Lett. A **378**, 178–183 (2014)

87. Li, C., Sprott, J.C., Yuan, Z., Li, H.: Constructing chaotic systems with total amplitude control. Int. J. Bifurc. Chaos **25**(1530), 025–14 (2015)

88. Li, G.H.: Inverse lag synchronization in chaotic systems. Chaos Solitons Fract. **40**, 1076–1080 (2009)

89. Liu, W., Qian, X., Yang, J., Xiao, J.: Antisynchronization in coupled chaotic oscillators. Phys. Lett. A **354**, 119–125 (2006)

90. Liu, X., Chen, T.: Synchronization of identical neural networks and other systems with an adaptive coupling strength. Int. J. Circuit Theory Appl. **38**, 631–648 (2010)

91. Lorenz, E.: Deterministic nonperiodic flow. J. Atmos. Sci. **20**, 130–141 (1963)

92. Mainieri, R., Rehacek, J.: Projective synchronization in three-dimensional chaotic system. Phys. Rev. Lett. **82**, 3042–3045 (1999)

93. Margaris, W.: Theory of the Non-linear Analog Phase Locked Loop. Springer, NJ (2004)

94. Maritan, A., Banavar, J.: Chaos noise and synchronization. Phys. Rev. Lett. **72**, 1451–1454 (1994)

95. Markus, L., Yamabe, H.: Global stability criteria for differential systems. Osaka Math. J. **12**, 305–317 (1960)

96. Mihajlovic, N., van Veggel, A., van de Wouw, N., Nijmeijer, H.: Analysis of friction-induced limit cycling in an experimental drill-string system. J. Dyn. Syst. Meas. Control **126**, 709–720 (2004)
97. Molaie, M., Jafari, S., Sprott, J.C., Golpayegani, S.M.R.H.: Simple chaotic flows with one stable equilibrium. Int. J. Bifurc. Chaos **23**, 1350,188 (2013)
98. Mosekilde, E., Postnov, D., Maistrenko, Y.: Chaotic Synchronization: Applications to Living Systems. World Scientific, Singapore (2002)
99. Nose, S.: A molecular dynamics method for simulations in the canonical ensemble. Mol. Phys. **52**, 255–268 (1984)
100. Ott, E.: Chaos in Dynamical Systems. Cambridge University Press, Cambridge (1993)
101. Ott, E., Grebogoi, C., Yorke, J.A.: Controlling chaos. Phys. Rev. Lett. **64**, 1196–1199 (1990)
102. Parlitz, U., Junge, L., Lauterborn, W., Kocarev, L.: Experimental observation of phase synchronization. Phys. Rev. E **54**, 2115–2217 (1996)
103. Pecora, L., Carroll, T.L.: Synchronization in chaotic systems. Phys. Rev. Lett. **64**, 821–824 (1990)
104. Petrovskii, I.G., Landis, Y.M.: On the number of limit cycles of the equation $dy/dx = P(x, y)/Q(x, y)$, where P and Q are 2–nd degree polynomials. Mat. Sb. (N.S.) **37**, 209–250 (1955)
105. Petrovskii, I.G., Landis, Y.M.: Corrections to the papers on the number of limit cycles of the equation $dy/dx = P(x, y)/Q(x, y)$, where P and Q are 2–nd degree polynomials and on the number of limiting cycles of the equation $dy/dx = P(x, y)/Q(x, y)$, where P and Q are polynomials. Mat. Sb. (N.S.) **48**, 253–255 (1959)
106. Pham, V.T., Jafari, S., Wang, X., Ma, J.: A chaotic system with different shapes of equilibria. Int. J. Bifurc. Chaos **26**, 1650,069 (2016)
107. Pham, V.T., Rahma, F., Frasca, M., Fortuna, L.: Families of transverse curves for two–dimensional systems of differential equations. Vestnik St. Petersburg University, pp. 48–78 (2006)
108. Pham, V.T., Rahma, F., Frasca, M., Fortuna, L.: Dynamics and synchronization of a novel hyperchaotic system without equilibrium. Int. J. Bifurc. Chaos **24**, 1450,087 (2014)
109. Pham, V.T., Volos, C.K., Vaidyanathan, S., Le, T.P., Vu, V.Y.: A memristor-based hyperchaotic system with hidden attractors: dynamics, synchronization and circuital emulating. J. Eng. Sci. Technol. Rev. **8**, 205–214 (2015)
110. Pikovski, A.S., Rabinovich, M.I., Trakhtengerts, V.Y.: Onset of stochasticity in decay confinement of parametric instability. Sov. Phys. JETP **47**, 715–719 (1978)
111. Pikovsky, A., Rosenblum, M., Kurths, J.: Synchronization: A Universal Concept in Nonlinear Sciences, 1st edn. Cambridge University Press, Cambridge (2003)
112. Pikovsky, A.S.: On the interaction of strange attractors. Z. Phys. B Condens. Matter **55**, 149–154 (1984)
113. Pliss, V.A.: Some Problems in the Theory of the Stability of Motion. Izd LGU, Leningrad (1958)
114. van der Pol, B.: On relaxation-oscillations. Philos. Mag. J. Sci. **7**, 978–992 (1926)
115. Poland, D.: Cooperative catalysis and chemical chaos: a chemical model for the Lorenz equations. Physica D **65**, 86–99 (1993)
116. Posch, H.A., Hoover, W.G., Vesely, F.J.: Canonical dynamics of the Nose oscillator: stability, order, and chaos. Phys. Rev. A **33**, 4253–4265 (1986)
117. Pyragas, K.: Continuous control of chaos by self-controlling feedback. Phys. Lett. A **170**, 421–428 (1992)
118. Pyragas, K., Lange, F., Letz, T., Parisi, J., Kittel, A.: Stabilization of an unstable steady state in intracavity frequency-doubled lasers. Phys. Rev. E **61**, 3721 (2000)
119. Rabinovich, M.I.: Stochastic autooscillations and turbulence. Uspehi Physicheskih **125**, 123–168 (1978)
120. Rayleigh, J.W.S.: The Theory of Sound. MacMillan, London (1877)
121. Rosenblum, M.G., Pikovsky, A.S., Kurths, J.: From phase to lag synchronization in coupled chaotic oscillators. Phys. Rev. Lett. **78**, 4193–4196 (1997)

122. Rulkov, N.F., Sushchik, M.M., Tsimring, L.S., Abarbanel, H.D.I.: Generalized synchronization of chaos in directionally coupled chaotic systems. Phys. Rev. E **51**, 980–994 (1995)
123. Sackellares, J.C., Iasemidis, L.D., Gilmore, R.L., Roper, S.N.: Chaos in the Brain?. World Scientific, Singapore (2000)
124. Schwartz, I.B., Carr, T.W., Triandaf, I.: Tracking controlled chaos: theoretical foundations and applications. Chaos **7**, 664–679 (1997)
125. Shahzad, M., Pham, V.T., Ahmad, M.A., Jafari, S., Hadaeghi, F.: Synchronization and circuit design of a chaotic system with coexisting hidden attractors. Eur. Phys. J. Spec. Top. **224**, 1637–1652 (2015)
126. Sharma, P.R., Shrimali, M.D., Prasad, A., Kuznetsov, N.V., Leonov, G.A.: Controlling dynamics of hidden attractors. Int. J. Bifurc. Chaos **25**, 1550,061 (2015)
127. Shi, S.: A concrete example of the existence of four limit cycles for plane quadratic systems. Sci. Sinica **23**, 153–158 (1980)
128. Sinha, S., Rao, J.S., Ramaswamy, R.: Adaptive control in nonlinear dynamics. Physica D **43**, 118–128 (1990)
129. Smale, S.: Mathematical problems for the next century. Math. Intell. **20**, 7–15 (1998)
130. Sommerfeld, A.: Beitrage zum dynamischen ausbau der festigkeitslehre. Zeitschrift des Vereins deutscher Ingenieure **4**, 391 (1902)
131. Sprott, J.: Some simple chaotic flows. Phys. Rev. E **50**, R647–650 (1994)
132. Sprott, J.C.: Strange attractors with various equilibrium types. Eur. Phys. J. Spec. Top. **224**, 1409–1419 (2015)
133. Sprott, J.C., Hoover, W.G., Hoover, C.G.: Heat conduction, and the lack thereof, in time–reversible dynamical systems: generalized Nose-Hoover oscillators with a temperature gradient. Phys. Rev. E **89**, 042,914 (2014)
134. Stoker, J.J.: Nonlinear Vibrations in Mechanical and Electrical Systems. Interscience, New York (1950)
135. Strogatz, S.H.: Nonlinear Dynamics and Chaos: With Applications to Physics, Biology, Chemistry, and Engineering. Westview Press, USA (1994)
136. Suarez, A., Quere, R.: Stability Analysis of Nonlinear Microwave Circuits. Artech House, NJ (2003)
137. Szatmari, I., Chua, L.O.: Awakening dynamics via passive coupling and synchronization mechanism in oscillatory cellular neural/nonlinear network. Int. J. Circuit Theory Appl. **36**, 525–553 (2008)
138. Taherion, S., Lai, Y.C.: Observability of lag synchronization of coupled chaotic oscillators. Phys. Rev. E **59**, R6247–R6250 (1999)
139. Timoshenko, S.: Vibration Problems in Engineering. Van Nostrand, New York (1928)
140. Tognoli, E., Kelso, J.A.S.: Brain coordination dynamics: true and false faces of phase synchrony and metastability. Prog. Neurobiol. **87**, 31–40 (2009)
141. Triandaf, I., Schwartz, I.B.: Tracking sustained chaos: a segmentation method. Phys. Rev. E **62**, 3529 (2000)
142. Tsuji, S., Ueta, T., Kawakami, H.: Bifurcation analysis of current coupled BVP oscillators. Int. J. Bifurc. Chaos **17**, 837–850 (2007)
143. Tzenov, S.I.: Strange attractors characterizing the osmotic instability, pp. 1–6 (2014). arXiv:1406.0979
144. Vaidyanathan, S., Pham, V.T., Volos, C.K.: A 5-D hyperchaotic Rikitake dynamo system with hidden attractors. Eur. Phys. J. Spec. Topics **224**, 1575–1592 (2015)
145. Vaidyanathan, S., Volos, C.K., Pham, V.T.: Analysis, control, synchronization and SPICE implementation of a novel 4-D hyperchaotic Rikitake dynamo system without equilibrium. J. Eng. Sci. Technol. Rev. **8**, 232–244 (2015)
146. Vembarasan, V., Balasubramaniam, P.: Chaotic synchronization of Rikitake system based on T-S fuzzy control techniques. Nonlinear Dyn. **74**, 31–44 (2013)
147. Vincent, U.E.: Synchronization of rikitake chaotic attractor using active control. Phys. Lett. A **343**, 133–138 (2005)
148. Viterbi, A.: Principles of Coherent Communications. McGraw-Hill, New York (1966)

149. Volos, C.K., Kyprianidis, I.M., Stouboulos, I.N.: Experimental demonstration of a chaotic cryptographic scheme. WSEAS Trans. Circuit Syst. **5**, 1654–1661 (2006)
150. Voss, H.U.: Anticipating chaotic synchronization. Phys. Rev. E **61**, 5115–5119 (2000)
151. Wang, J., Che, Y.Q., Zhou, S.S., Deng, B.: Unidirectional synchronization of Hodgkin-Huxley neurons exposed to ELF electric field. Chaos Solitons Fract. **39**, 1335–1345 (2009)
152. Wang, L., Yang, X.S.: The invariant tori of knot type and the interlinked invariant tori in the Nose-Hoover oscillator. Eur. Phys. J. B **88**, 1–5 (2015)
153. Wang, X., Chen, G.: A chaotic system with only one stable equilibrium. Commun. Nonlinear Sci. Numer. Simul. **17**, 1264–1272 (2012)
154. Wang, X., Chen, G.: Constructing a chaotic system with any number of equilibria. Nonlinear Dyn. **71**, 429–436 (2013)
155. Wang, Z., Cang, S., Ochola, E., Sun, Y.: A hyperchaotic system without equilibrium. Nonlinear Dyn. **69**, 531–537 (2012)
156. Wang, Z., Ma, J., Cang, S., Wang, Z., Chen, Z.: Simplified hyper-chaotic systems generating multi-wing non-equilibrium attractors. Optik **127**, 2424–2431 (2016)
157. Wei, Z.: Dynamical behaviors of a chaotic system with no equilibria. Phys. Lett. A **376**, 102–108 (2011)
158. Wei, Z., Moroz, I., Liu, A.: Degenerate Hopf bifurcation, hidden attractors, and control in the extented Sprott E system with only one stable equilibrium. Turk. J. Math. **38**, 672–687 (2014)
159. Wei, Z., Wang, R., Liu, A.: A new finding of the existence of hidden hyperchaotic attractors with no equilibria. Math. Comput. Simul. **100**, 13–23 (2014)
160. Wei, Z., Wang, Z.: Chaotic behavior and modified function projective synchronization of a simple system with one stable equilibrium. Kybernetika **49**, 359–374 (2013)
161. Wembe, E.T., Yamapi, R.: Chaos synchronization of resistively coupled Duffing systems: numerical and experimental investigations. Commun. Nonlinear Sci. Numer. Simul. **14**, 1439–1453 (2009)
162. Wendt, K., Fredentall, G.: Automatic frequency and phase control of synchronization in TV receivers. Proc. IRE **31**, 1–15 (1943)
163. Woafo, P., Kadji, H.G.E.: Synchronized states in a ring of mutually coupled self–sustained electrical oscillators. Phys. Rev. E **69**, 046,206 (2004)
164. Wu, C.W.: Synchronization in Coupled Chaotic Circuits and System, 1st edn. World Scientific, Singapore (2002)
165. Wu, X.J., Liu, J.S., Chen, G.R.: Chaos synchronization of Rikitake chaotic attractor using the passive control technique. Nonlinear Dyn. **53**, 45–53 (2008)
166. Yang, T., Chua, L.O.: Control of chaos using sampled-data feedback control. Int. J. Bifurc. Chaos **8**, 2433–2438 (1998)
167. Yassen, M.T.: Chaos control of chaotic dynamical systems using backstepping design. Chaos Solitions Fract. **27**, 537–548 (2006)
168. Yau, H.T., Yan, J.J.: Design of sliding mode controller for lorenz chaotic system with nonlinear input. Chaos Solitions Fract. **19**, 891–898 (2004)
169. Zhang, H., Liu, D., Wang, Z.: Controlling Chaos: Suppression. Synchronization and Chaotification. Springer, Germany (2009)
170. Zhong, G.Q., Man, K.F., Ko, K.T.: Uncertainty in chaos synchronization. Int. J. Bifurc. Chaos **11**, 1723–1735 (2001)
171. Zuo, J., Li, C.: Multiple attractors and dynamic analysis of a no-equilibrium chaotic system. Optik **127**, 7952–7959 (2016)

Chapter 2
Systems with Stable Equilibria

2.1 Wang–Chen System with Only One Stable Equilibrium

Sprott has proposed 19 remarkable flows with quadratic nonlinearities [20]. In the list of Sprott's systems, the Sprott E system is described by

$$\begin{cases} \dot{x} = yz \\ \dot{y} = x^2 - y \\ \dot{z} = 1 - 4x \end{cases} \tag{2.1}$$

The Sprott E system has five terms and two nonlinearities. Interestingly, Sprott E system (2.1) has a degenerate equilibrium $E(0.25, 0.0625, 0)$ with eigenvalues $\lambda_1 = -1$ and $\lambda_{2,3} = \pm 0.5i$. By adding a control parameter a to the Sprott E system, Wang and Chen introduced a new special system [27]

$$\begin{cases} \dot{x} = yz + a \\ \dot{y} = x^2 - y \\ \dot{z} = 1 - 4x \end{cases} \tag{2.2}$$

with only one equilibrium point $E(0.25, 0.0625, -16a)$. System (2.2) is a three-dimensional autonomous one with only quadratic nonlinearities. It is easy to see that Wang–Chen system is dissipative because

$$\nabla V = \frac{\partial \dot{x}}{\partial x} + \frac{\partial \dot{y}}{\partial y} + \frac{\partial \dot{z}}{\partial z} = -1 < 0 \tag{2.3}$$

By linearizing Wang–Chen system at the equilibrium point E, the Jacobian matrix is given by

$$\mathbf{J}_E = \begin{bmatrix} 0 & -16a & 0.0625 \\ 0.5 & -1 & 0 \\ -4 & 0 & 0 \end{bmatrix} \tag{2.4}$$

© The Author(s) 2017
V.-T. Pham et al., *Systems with Hidden Attractors*,
SpringerBriefs in Nonlinear Circuits, DOI 10.1007/978-3-319-53721-4_2

Fig. 2.1 Phase portrait in the x–y plane of Wang–Chen system with only one stable equilibrium for $a = 0.006$ and initial conditions $(x(0), y(0), z(0)) = (1, 1, 1)$

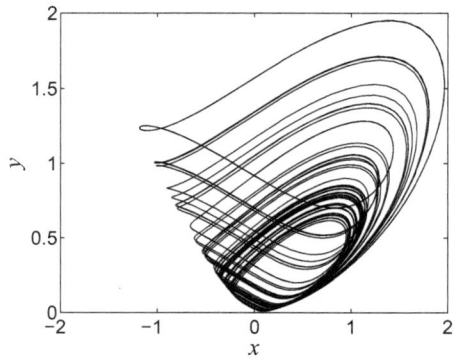

Therefore, the characteristic equation of Wang–Chen system is

$$\lambda^3 + \lambda^2 + (0.25 + 8a)\,\lambda + 0.25 = 0 \tag{2.5}$$

According to Routh–Hurwitz criterion, the equilibrium is stable for

$$\begin{cases} 0.25 + 8a > 0 \\ 1\,(0.25 + 8a) > 0.25 \end{cases} \tag{2.6}$$

that is $a > 0$. Interestingly, Wang–Chen system with only one stable equilibrium generates chaotic behavior [27]. For example, the phase portrait of Wang–Chen system for $a = 0.006$ is shown in Fig. 2.1. In this case, Wang–Chen system has only one stable equilibrium $E(0.25, 0.0625, -0.096)$ with three eigenvalues

$$\lambda_1 = -0.9607, \quad \lambda_{2,3} = -0.0197 \pm 0.5098i \tag{2.7}$$

Remarkably, the state variable z appears only in the first equation of Wang–Chen system (2.2). As a result, the state variable z is controllable by replacing z with $z + k_z$. Here, k_z is a control parameter. Hence, Wang–Chen system is rewritten in the following form

$$\begin{cases} \dot{x} = y\,(z + k_z) + a \\ \dot{y} = x^2 - y \\ \dot{z} = 1 - 4x \end{cases} \tag{2.8}$$

Obviously, we can vary the level of amplitude z by changing the value of the control parameter k_z.

Wang–Chen system is an attractive example for discovering mysterious features of chaos. By using a rigorous computer-aided approach, horseshoe chaos in Wang–Chen system was confirmed [9]. Wei and Wang introduced an extended version of Wang–Chen system

$$\begin{cases} \dot{x} = yz + h(x) \\ \dot{y} = x^2 - y \\ \dot{z} = 1 - 4x \end{cases} \tag{2.9}$$

where $h(x) = ex^2 + fx + g$ and e, f, and g are real parameters. Modified function projective synchronization between the extended system (2.9) and the Sprott E system (2.1) was reported in [32]. Degenerate Hopf bifurcations and adaptive control in the extended system (2.9) were also studied [29]. The coexistence of point, periodic and strange attractors in Wang–Chen system was presented by Sprott et al. [21]. Moreover, the effect of multiple delays on Wang–Chen system was analyzed by considering the stability of equilibrium and the existence of Hopf bifurcations [31]. Wang–Chen system with time-delayed controlling forces was designed as

$$\begin{cases} \dot{x} = yz + a + k_1 x(t - \tau_1) + k_2 x(t - \tau_2) \\ \dot{y} = x^2 - y \\ \dot{z} = 1 - 4x \end{cases} \tag{2.10}$$

in which k_1 and k_2 are the gains of the time delays τ_1 and τ_2, respectively [31]. The striking discovery of Wang–Chen system is a key work motivating others researches.

2.2 Simple Flows with One Stable Equilibrium

Following up the discovery of Wang–Chen system, Molaie et al. performed an effective approach to find three-dimensional autonomous flows with one stable equilibrium [17]. The approach includes three main steps: (a) setting a general equation, (b) searching potential cases, and (c) selecting elegant cases.

In the first step, a general equation with quadratic nonlinearities and unknown coefficients is constructed. It is noteworthy that the general system must have only one equilibrium. In the second step, a computer search routine is implemented to find various potential cases with different coefficients. In the last step, the most elegant cases are selected. An elegant case is a system with as many coefficients as possible are set to zero while the others are ± 1 or set to a small integer or decimal fraction with the fewest digits. Molaie et al. reported 23 simple chaotic systems with a stable equilibrium [17].

In order to illustrate such interesting approach, let us consider an example. We consider a general equation with quadratic nonlinearities with nine coefficients $(a_1 - a_9)$ as follows:

$$\begin{cases} \dot{x} = y \\ \dot{y} = -x + yz \\ \dot{z} = a_1 x + a_2 y + a_3 z + a_4 x^2 + a_5 y^2 + a_6 xy + a_7 xz + a_8 yz + a_9 \end{cases} \tag{2.11}$$

It is noteworthy that the general form (2.11) has only one equilibrium point

$$E\left(0, 0, -\frac{a_9}{a_3}\right) \tag{2.12}$$

By linearizing the general form at the equilibrium point E, the Jacobian matrix is given by:

$$\mathbf{J}_E = \begin{bmatrix} 0 & 1 & 0 \\ -1 & -\frac{a_9}{a_3} & 0 \\ a_1 - \frac{a_7 a_9}{a_3} & a_2 & -\frac{a_8 a_9}{a_3} \ a_3 \end{bmatrix} \tag{2.13}$$

Therefore, the characteristic equation of the system is

$$\lambda^3 + \left(\frac{a_9}{a_3} - a_3\right)\lambda^2 + (1 - a_9)\lambda - a_3 = 0 \tag{2.14}$$

According to the Routh–Hurwitz criterion, the equilibrium is stable for

$$\begin{cases} \frac{a_9}{a_3} - a_3 > 0 \\ 1 - a_9 > 0 \\ -a_3 > 0 \\ \left(\frac{a_9}{a_3} - a_3\right)(1 - a_9) > -a_3 \end{cases} \tag{2.15}$$

By applying the search routine to the general form (2.11) with the condition (2.15), many systems are found, for example seven systems from SE_7 to SE_{13} [17]. Remarkably, one of the most elegant systems is the Molaie SE_8 system [17] for

$$\begin{cases} a_1 = a_2 = a_6 = a_7 = a_8 = 0 \\ a_3 = -1, a_5 = 1 \\ a_4 = -0.7, a_9 = -0.1 \end{cases} \tag{2.16}$$

The Molaie SE_8 system is described by

$$\begin{cases} \dot{x} = y \\ \dot{y} = -x + yz \\ \dot{z} = -z - 0.7x^2 + y^2 - 0.1 \end{cases} \tag{2.17}$$

Substituting (2.16) into (2.12) and (2.14), we obtain the equilibrium of the Molaie SE_8 system

$$E\,(0, 0, 0.1) \tag{2.18}$$

while the three eigenvalues are

$$\lambda_1 = -1.0000, \lambda_{2,3} = -0.0500 \pm 0.9987i \tag{2.19}$$

Fig. 2.2 Phase portrait in the x–y plane of the Molaie SE$_8$ system with only one stable equilibrium for initial conditions $(x(0), y(0), z(0)) = (0, 0.9, 0)$

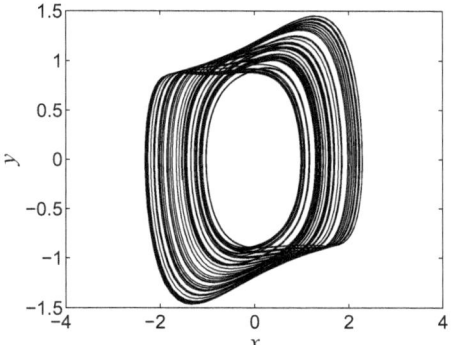

From (2.19), it is easy to verify that the equilibrium E is stable. However, the Molaie SE$_8$ system can exhibit complex behavior as illustrated in Fig. 2.2.

Similarly, we can study Lao's system [13] and Kingni's system [12] by considering another general equation with quadratic nonlinearities with nine coefficients (a_1–a_9)

$$\begin{cases} \dot{x} = -z \\ \dot{y} = -x - z \\ \dot{z} = a_1 x + a_2 y + a_3 z + a_4 x^2 + a_5 z^2 + a_6 xy + a_7 xz + a_8 yz + a_9 \end{cases} \tag{2.20}$$

in which the equilibrium is

$$E\left(0, -\frac{a_9}{a_2}, 0\right) \tag{2.21}$$

Lao system is a dynamical system with nine terms [13]

$$\begin{cases} \dot{x} = -z \\ \dot{y} = -x - z \\ \dot{z} = 2x - 1.3y - 2z + x^2 + z^2 - xz \end{cases} \tag{2.22}$$

Lao system has only one stable equilibrium $E(0, 0, 0)$, with three eigenvalues

$$\lambda_1 = -1.9783, \lambda_{2,3} = -0.0108 \pm 0.8106i \tag{2.23}$$

Authors developed a new parameter estimation procedure for the system (2.22) via a cost function based on Gaussian Mixture Model [13].

In addition, Kingni introduced a three-dimensional autonomous system

$$\begin{cases} \dot{x} = -z \\ \dot{y} = -x - z \\ \dot{z} = 3x - 1.3y + x^2 - z^2 - yz + 1.01 \end{cases} \tag{2.24}$$

Kingni system has only one stable equilibrium $E\left(0, \frac{101}{130}, 0\right)$ with three eigenvalues

$$\lambda_1 = -0.7679, \lambda_{2,3} = -0.0045 \pm 1.3012i \tag{2.25}$$

Dynamics of the Kingni system were described via numerical simulations such as phase portraits, bifurcation diagrams, and new cost function for parameter estimation [13]. In addition, Wei et al. studied theoretically some complex dynamics of the Kingni system using the Poincaré compactification of polynomial vector, degenerate Hopf bifurcation, and zero-Hopf bifurcation [30].

2.3   Systems with Stable Equilibrium Points

In 2008, Yang and Chen introduced a three-dimensional system described by

$$\begin{cases} \dot{x} = a\,(y - x) \\ \dot{y} = cx - xz \\ \dot{z} = -bz + xy \end{cases} \tag{2.26}$$

in which a, b, and c are parameters [38]. Yang–Chen system has three equilibrium points

$$\begin{cases} E_1\,(0, 0, 0) \\ E_2\left(\sqrt{bc}, \sqrt{bc}, c\right) \\ E_3\left(-\sqrt{bc}, -\sqrt{bc}, c\right) \end{cases} \tag{2.27}$$

Yang–Chen system is topologically different from well-known systems such as Lorenz system, Chen system, and Lü system [38].

For $a = c = 35, b = 3$, Yang–Chen system (2.26) becomes

$$\begin{cases} \dot{x} = 35\,(y - x) \\ \dot{y} = 35x - xz \\ \dot{z} = -3z + xy \end{cases} \tag{2.28}$$

with three equilibrium points

$$\begin{cases} E_1\,(0, 0, 0) \\ E_2\left(\sqrt{105}, \sqrt{105}, 35\right) \\ E_3\left(-\sqrt{105}, -\sqrt{105}, 35\right) \end{cases} \tag{2.29}$$

Fig. 2.3 Phase portrait in the
$x–y$ plane of the Yang–Chen
system with one saddle and
two stable node-foci for
initial conditions
$(x(0), y(0), z(0)) =$
$(1.15, 3.5, 3.3)$

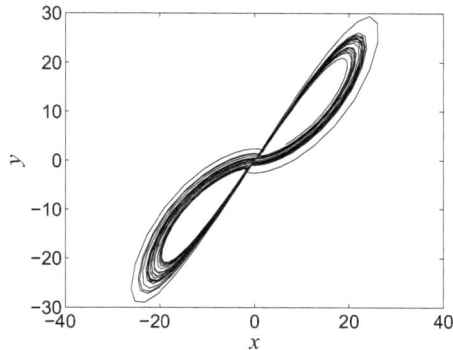

The eigenvalues at the equilibrium E_1 are

$$\lambda_1 = -3, \lambda_2 = -\frac{35\left(\sqrt{5}+1\right)}{2}, \lambda_3 = \frac{35\left(\sqrt{5}-1\right)}{2} \tag{2.30}$$

while the eigenvalues at the equilibrium points $E_{2,3}$ are

$$\lambda_1 = -37.6122, \lambda_{2,3} = -0.1939 \pm 13.9778i \tag{2.31}$$

It is interesting that the equilibrium E_1 is a saddle, and the equilibrium points $E_{2,3}$ are two stable node-foci. Moreover, Yang–Chen system (2.28) exhibits complex dynamics (see Fig. 2.3).

It is natural to ask a question whether there exist systems with only stable node-foci. There are a few positive answers for this attractive question [33, 34, 39]. A typical example of such systems was proposed by Yang et al. [39]

$$\begin{cases} \dot{x} = a\,(y - x) \\ \dot{y} = -cy - xz \\ \dot{z} = xy - b \end{cases} \tag{2.32}$$

in which a and b are two parameters. There are two equilibrium points in system (2.32)

$$\begin{cases} E_1\left(\sqrt{b}, \sqrt{b}, -c\right) \\ E_2\left(-\sqrt{b}, -\sqrt{b}, -c\right) \end{cases} \tag{2.33}$$

For $a = 10, b = 100$, and $c = 11.2$, Yang system (2.32) becomes

$$\begin{cases} \dot{x} = 10\,(y - x) \\ \dot{y} = -11.2y - xz \\ \dot{z} = xy - 100 \end{cases} \tag{2.34}$$

with two equilibrium points

$$\begin{cases} E_1 \, (10, \, 10, \, -11.2) \\ E_2 \, (-10, \, -10, \, -11.2) \end{cases} \tag{2.35}$$

Interestingly, the equilibrium points E_1 and E_2 are the two stable node-foci because the eigenvalues at the equilibrium points are

$$\lambda_1 = -20.9778, \, \lambda_{2,3} = -0.1111 \pm 9.7635i \tag{2.36}$$

Yang system (2.32) contains different special cases such as diffusionless Lorenz system and Burke–Shaw system. The Hopf bifurcation and singularly degenerate heteroclinic and homoclinic orbits in Yang system were analyzed [39]. Yang system with time-delay feedback [28] was designed as

$$\begin{cases} \dot{x} = a \, (y - x) \\ \dot{y} = -cy - xz + k \, (y \, (t - \tau) - y) \\ \dot{z} = xy - b \end{cases} \tag{2.37}$$

in which τ is the time delay and k indicates the gain of the time-delay feedback. Chaos in system (2.37) could be controlled by time-delay feedback [28].

By generalizing the Sprott C system, a system [34] was constructed as

$$\begin{cases} \dot{x} = a \, (y - x) \\ \dot{y} = -cy - xz \\ \dot{z} = y^2 - b \end{cases} \tag{2.38}$$

in which a and b are parameters. System (2.38) possesses two equilibria

$$\begin{cases} E_1 \left(\sqrt{b}, \, \sqrt{b}, \, -c \right) \\ E_2 \left(-\sqrt{b}, \, -\sqrt{b}, \, -c \right) \end{cases} \tag{2.39}$$

For $a = 10, b = 100$, and $c = 0.4$, the eigenvalues at the equilibrium points E_1, E_2 are

$$\lambda_1 = -10.1357, \, \lambda_{2,3} = -0.1321 \pm 14.0465i \tag{2.40}$$

As a result, system (2.38) has two stable equilibria. Wei and Pehlivan also investigated the coexisting attractors and the circuit design of system with only two stable equilibria (2.38) in [35].

Another interesting system with an exponential nonlinear term [33] should be mentioned

$$\begin{cases} \dot{x} = a \, (y - x) \\ \dot{y} = -by + mxz \\ \dot{z} = n - e^{xy} \end{cases} \tag{2.41}$$

where a, b, m, n are parameters. The presence of an exponential nonlinear term makes the system (2.41) different from the systems (2.32) and (2.38). It is easy to get two equilibrium points in system (2.41)

$$\begin{cases} E_1\left(\sqrt{\ln n}, \sqrt{\ln n}, \frac{b}{m}\right) \\ E_2\left(-\sqrt{\ln n}, -\sqrt{\ln n}, \frac{b}{m}\right) \end{cases} \tag{2.42}$$

For $a = 0.8696$, $b = 2.1$, $m = 0.756144$, and $n = 10.5$, the eigenvalues at such equilibrium points E_1, E_2 are

$$\lambda_1 = -1.9468, \lambda_{2,3} = -0.5114 \pm 4.0517i \tag{2.43}$$

Obviously, E_1 and E_2 are stable equilibria.

2.4 Constructing a System with One Stable Equilibrium

Although the fact that different systems with stable equilibrium have been reported, how we can achieve a new system with stable equilibrium from another known one is still an attractive topic. To the best of our knowledge, there are a few discussions about this topic in the literature [19]. In this section, we consider a new system with only one stable equilibrium which is constructed from a known system with infinite equilibrium.

We start with the chaotic flow LE_1 which was introduced by Jafari and Sprott in [10]

$$\begin{cases} \dot{x} = y \\ \dot{y} = -x + yz \\ \dot{z} = -x - axy - bxz \end{cases} \tag{2.44}$$

in which a and b are two positive parameters. It is trivial to see that the LE_1 system has infinite equilibria

$$E\,(0, 0, z) \tag{2.45}$$

By adding a positive control parameter c to the LE_1 system, we obtain a new system

$$\begin{cases} \dot{x} = y \\ \dot{y} = -x + yz + c \\ \dot{z} = -x - axy - bxz \end{cases} \tag{2.46}$$

The system (2.46) has only one equilibrium

$$E\left(c, 0, -\frac{1}{b}\right) \tag{2.47}$$

The Jacobian matrix at the equilibrium E is determined as

$$\mathbf{J}_E = \begin{bmatrix} 0 & 1 & 0 \\ -1 & -\frac{1}{b} & 0 \\ 0 & -ac & -bc \end{bmatrix} \tag{2.48}$$

and its characteristic equation is

$$\lambda^3 + \left(bc + \frac{1}{b}\right)\lambda^2 + (c+1)\lambda + bc = 0 \tag{2.49}$$

According to the Routh–Hurwitz stability criterion, the equilibrium is stable because

$$\begin{cases} bc + \frac{1}{b} > 0 \\ c + 1 > 0 \\ bc > 0 \\ \left(bc + \frac{1}{b}\right)(c+1) > bc \end{cases} \tag{2.50}$$

From Eq. (2.49), the eigenvalues are

$$\lambda_1 = -bc, \lambda_{2,3} = -\frac{1}{2b} \pm i\frac{1}{2b}\sqrt{4b^2 - 1} \tag{2.51}$$

For $a = 15$, $b = 1$, and $c = 0.001$, eigenvalues of system (2.46) at the equilibrium $E(0.001, 0, -1)$ are

$$\lambda_1 = -0.001, \lambda_{2,3} = -0.5 \pm 0.866i \tag{2.52}$$

Interestingly, although system (2.46) has only one stable equilibrium, it can exhibit chaotic attractor as shown in Fig. 2.4. The system is a typical example to show how we can construct a new system with only one stable equilibrium from another system with an infinite number of equilibrium.

Fig. 2.4 Phase portrait in the x–y plane of the system with only one stable equilibrium (2.46) for $a = 15, b = 1, c = 0.001$, and initial conditions $(x(0), y(0), z(0)) = (0, 0.5, 0.5)$

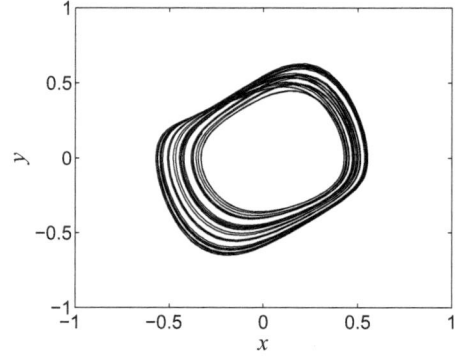

2.5 Double-Scroll Attractors in Systems with Stable Equilibria

It has been known that systems with multi-scroll attractors generate more complex dynamics compared with normal systems with few attractors [15]. As a result, systems with multi-scroll attractors have been considered as potential candidates for using in chaos-based applications. Several attempts have been made to encrypt fingerprint images through a two-dimensional chaotic sequence achieved from multi-scroll chaotic attractors [6]. By increasing the number of scrolls in generalized Jerk circuit, the entropy of a random number generator was improved [37]. Gamez-Guzan et al. transmitted secure information by considering the synchronization of Chua's circuits with multi-scroll attractor [3]. In addition, Orue et al. applied a parameter determination method for double-scroll chaotic systems into chaotic cryptanalysis [18]. The issue of generating multi-scroll chaotic attractors has received considerable critical attention [16, 25, 26, 41, 42]. When studying systems with hidden attractors, a naturally attractive question is posed that "Can systems with hidden attractors exhibit multi-scroll attractors?"

It has previously been observed that equilibrium points of dynamical systems play a vital role when designing multi-scroll attractors [15]. For instance, by increasing the number of unstable equilibrium points in a known system, we can get multi-scroll attractors [16, 42]. However, it is interesting that mentioned systems with two equilibria in previous sections can generate double-scroll attractors. For example, double-scroll attractor of system (2.38) are illustrated in Fig. 2.5. Therefore, the relationships between the equilibria and their properties with the strange attractor in dynamical systems are quite subtle and are still needed to be studied further.

Fig. 2.5 Double-scroll attractor in the x–y plane of the system with two stable equilibrium points (2.38) for $a = 10, b = 100, c = 0.4$ and initial conditions $(x(0), y(0), z(0)) = (12.2, 4.81, -0.2)$

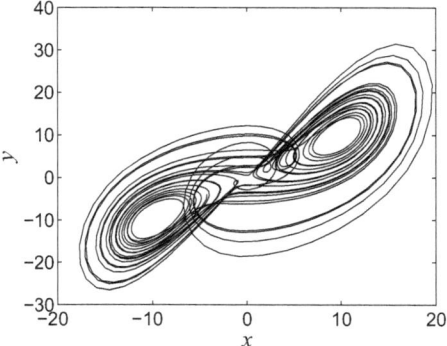

2.6 Fractional-Order Form of a System with Stable Equilibrium

As it is known that practical models such as heat conduction, electrode–electrolyte polarization, electronic capacitors, dielectric polarization, viscoelastic systems are more adequately described by the fractional-order different equations [1, 8, 11, 22, 24, 36]. Adams–Bashforth–Mounlton numerical algorithm is often used to investigate fractional-order differential equations [2, 4, 23]. Here, we present this algorithm briefly.

We consider the fractional-order differential equation as follows:

$$\begin{cases} \frac{d^q x(t)}{dt^q} = f\left(t, x\left(t\right)\right), & 0 \le t \le T, \\ x^{(i)}\left(0\right) = x_0^{(i)} & i = 0, 1, \dots, m - 1, \end{cases} \tag{2.53}$$

where $m - 1 < q \le m \in Z^+$. Equation (2.53) is equivalent to the following Volterra integral equation:

$$x\left(t\right) = \sum_{i=0}^{m-1} \frac{t^i}{i!} x_0^{(i)} + \frac{1}{\Gamma\left(q\right)} \int_0^t \left(t - \tau\right)^{q-1} f\left(\tau, x\left(\tau\right)\right) d\tau, \tag{2.54}$$

in which the Gamma function $\Gamma\left(.\right)$ is defined as

$$\Gamma\left(q\right) = \int_0^\infty e^{-t} t^{q-1} dt. \tag{2.55}$$

We set $h = \frac{T}{N}$, $N \in Z^+$, and $t_n = nh$ $(n = 0, 1, \dots, N)$. So we can discrete Eq. (2.54) as follows:

$$\begin{aligned} x_h\left(t_{n+1}\right) = & \sum_{i=0}^{m-1} \frac{t_{n+1}^i}{i!} x_0^{(i)} + \frac{h^q}{\Gamma(q+2)} f\left(t_{n+1}, x_h^p\left(t_{n+1}\right)\right) \\ & + \frac{h^q}{\Gamma(q+2)} \sum_{j=0}^n a_{j,n+1} f\left(t_j, x_h\left(t_j\right)\right), \end{aligned} \tag{2.56}$$

where

$$a_{j,n+1} = \begin{cases} n^{q+1} - (n - q)(n + 1)^q, & \text{if } j = 0, \\ (n - j + 2)^{q+1} + (n - j)^{q+1} \\ -2(n - j + 1)^{q+1}, & \text{if } 1 \le j \le n, \\ 1, & \text{if } j = n + 1. \end{cases} \tag{2.57}$$

It is noteworthy that the predicted value $x_h^p(t_{n+1})$ is calculated as

$$x_h^p(t_{n+1}) = \sum_{i=0}^{m-1} \frac{t_{n+1}^i}{i!} x_0^{(i)} + \frac{1}{\Gamma(q)} \sum_{j=0}^{n} b_{j,n+1} f\left(t_j, x_h(t_j)\right), \qquad (2.58)$$

in which

$$b_{j,n+1} = \frac{h^q}{q} \left((n+1-j)^q - (n-j)^q\right), \quad 0 \le j \le n. \qquad (2.59)$$

Here, the estimation error e in the method is given by:

$$e = \max \left| x(t_j) - x_h(t_j) \right| = O(h^p) \quad (j = 0, 1, \dots, N), \qquad (2.60)$$

with $p = \min(2, 1 + q)$.

Existence of chaos in fractional-order systems is investigated [5, 7, 14, 40]. Kingni et al. replaced integer-order derivatives in the system (2.24) with fractional-order ones [12]. As a result, authors described the fractional-order system with only one stable equilibrium by

$$\begin{cases} \dfrac{d^{\alpha_1} x}{dt^{\alpha_1}} = -z \\[2mm] \dfrac{d^{\alpha_2} x}{dt^{\alpha_2}} = -x - z \\[2mm] \dfrac{d^{\alpha_3} x}{dt^{\alpha_3}} = 3x - ay + x^2 - z^2 - yz + b \end{cases} \qquad (2.61)$$

in which $\alpha_1, \alpha_2, \alpha_3$ are the derivative orders ($0 < \alpha_1, \alpha_2, \alpha_3 < 1$).

It is interesting that the fractional-order system (2.61) can generate chaos with both commensurate fractional orders and incommensurate fractional orders [12]. By applying Adams–Bashforth–Moulnton numerical algorithm [2, 4, 23], the phase portrait of the fractional-order system (2.61) with commensurate fractional orders ($\alpha_1 = \alpha_2 = \alpha_3 = \alpha = 0.999$) is depicted in Fig. 2.6.

Fig. 2.6 Chaotic attractor in the x–y plane of the fractional-order system with only one stable equilibrium (2.61) for commensurate fractional orders $\alpha_1 = \alpha_2 = \alpha_3 = \alpha = 0.999$ and initial conditions $(x(0), y(0), z(0)) = (0.7, 2.3, -1.5)$

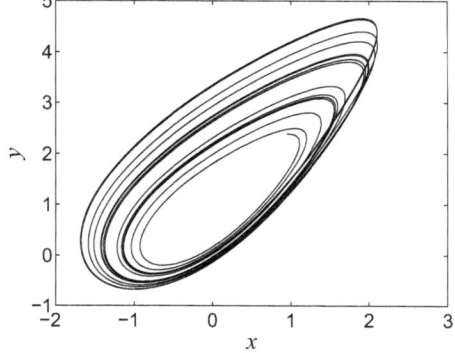

References

1. Bagley, R.L., Calico, R.A.: Fractional-order state equations for the control of visco-elastically damped structers. J. Guide Control Dyn. **14**, 304–311 (1991)
2. Diethelm, K., Ford, N.J., Freed, A.D.: A predictor-corrector approach for the numerical solution of fractional differential equations. Nonlinear Dyn. **29**, 3–22 (2002)
3. Gamez-Guzman, L., Cruz-Hernandez, C., Lopez-Gutierrez, R., Garcia-Guerrero, E.E.: Synchronization of Chua's circuits with multi-scroll attractors: application to communication. Commun. Nonlinear Sci. Numer. Simul. **14**, 2765–2775 (2009)
4. Gejji, D., Jafari, H.: Adomian decomposition: a tool for solving a system of fractional differential equations. J. Math. Anal. Appl. **301**, 508–518 (2005)
5. Grigorenko, I., Grigorenko, E.: Chaotic dynamics of the fractional–order lorenz system. Phys. Rev. Lett. **91**, 034,101 (2003)
6. Han, F., Hu, J., Yu, X., Wang, Y.: Fingerprint images encryption via multi-scroll chaotic attractors. Appl. Math. Comput. **185**, 931–939 (2007)
7. Hartley, T.T., Lorenzo, C.F., Qammer, H.K.: Chaos on a fractional Chua's system. IEEE Trans. Circ. Syst. I Fundam. Theory Appl. **42**, 485–490 (1995)
8. Heaviside, O.: Electromagnetic Theory. Academic Press, New York (1971)
9. Huan, S., Li, Q., Yang, X.: Horseshoes in a chaotic system with only one stable equilibrium. Int. J. Bifurc. Chaos **23**, 1350,002 (2013)
10. Jafari, S., Sprott, J.C.: Simple chaotic flows with a line equilibrium. Chaos, Solitons Fract. **57**, 79–84 (2013)
11. Jenson, V.G., Jeffreys, G.V.: Mathematical Methods in Chemical Engineering. Academic Press, New York (1997)
12. Kingni, S.T., Jafari, S., Simo, H., Woafo, P.: Three-dimensional chaotic autonomous system with only one stable equilibrium: analysis, circuit design, parameter estimation, control, synchronization and its fractional-order form. Eur. Phys. J. Plus **129**, 76 (2014)
13. Lao, S.K., Shekofteh, Y., Jafari, S., Sprott, J.C.: Cost function based on Gaussian mixture model for parameter estimation of a chaotic circuit with a hidden attractor. Int. J. Bifurc. Chaos **24**, 1450,010 (2014)
14. Li, C.P., Peng, G.J.: Chaos in Chen's system with a fractional-order. Chaos, Solitons Fract. **20**, 443–450 (2004)
15. Lü, J.H., Chen, G.R.: Generating multiscroll chaotic attractors: theories, methods and applications. Int. J. Bifurc. Chaos **16**, 775–858 (2006)
16. Ma, J., Wu, X., Chu, R., Zhang, L.: Selection of multi-scroll attractors in Jerk circuits and their verification using Pspice. Nonlinear Dyn. **76**, 1951–1962 (2014)
17. Molaie, M., Jafari, S., Sprott, J.C., Golpayegani, S.M.R.H.: Simple chaotic flows with one stable equilibrium. Int. J. Bifurc. Chaos **23**, 1350,188 (2013)
18. Orue, A.B., Alvarez, G., Pastor, G., Romera, M., Montoya, F., Li, S.: A new parameter determination method for some double-scroll chaotic systems and its applications to chaotic cryptanalysis. Commun. Nonlinear Sci. Numer. Simul. **15**, 3471–3483 (2010)
19. Pham, V.T., Volos, C., Jafari, S., Wei, Z., Wang, X.: Constructing a novel no–equilibrium chaotic system. Int. J. Bifurc. Chaos **24**, 1450,073 (2014)
20. Sprott, J.: Some simple chaotic flows. Phys. Rev. E **50**, R647–650 (1994)
21. Sprott, J.C., Wang, X., Chen, G.: Coexistence of point, periodic and strange attractors. Int. J. Bifurc. Chaos **23**, 1350,093 (2013)
22. Sun, H.H., Abdelwahad, A.A., Onaral, B.: Linear approximation of transfer function with a pole of fractional-order. IEEE Trans. Autom. Control **29**, 441–444 (1894)
23. Tavazoei, M.S., Haeri, M.: Limitations of frequency domain approximation for detecting chaos in fractional-order systems. Nonlinear Anal. **69**, 1299–1320 (2008)
24. Tavazoei, M.S., Haeri, M.: A proof for non existence of periodic solutions in time invariant fractional-order systems. Automatica **45**, 1886–1890 (2009)

25. Tlelo-Cuautle, E., Pano-Azucena, A.D., Rangel-Magdaleno, J.J., Carbajal-Gomez, V.H., Rodriguez-Gomez, G.: Generating a 50-scroll chaotic attractor at 66 MHz by using FPGAs. Nonlinear Dyn. **85**, 2143–2157 (2016)
26. Tlelo-Cuautle, E., Rangel-Magdaleno, J.J., Pano-Azucena, A.D., Obeso-Rodelo, P.J., Nunez-Perez, J.C.: FPGA realization of multi-scroll chaotic oscillators. Commun. Nonlinear Sci. Numer. Simul. **27**, 66–80 (2015)
27. Wang, X., Chen, G.: A chaotic system with only one stable equilibrium. Commun. Nonlinear Sci. Numer. Simul. **17**, 1264–1272 (2012)
28. Wei, Z.: Delayed feedback on the 3-D chaotic system only with two stable node-foci. Comput. Math. Appl. **63**, 728–738 (2012)
29. Wei, Z., Moroz, I., Liu, A.: Degenerate Hopf bifurcation, hidden attractors, and control in the extented Sprott E system with only one stable equilibrium. Turk. J. Math. **38**, 672–687 (2014)
30. Wei, Z., Moroz, I., Wang, Z., Sprott, J.C., Kapitaniak, T.: Dynamics at infinity, degenerate Hopf and zero–Hopf bifurcation for Kingni–Jafari system with hidden attractors. Int. J. Bifurc. Chaos **26**, 1650,125 (2016)
31. Wei, Z., Pham, V.T., Kapitaniak, T., Wang, Z.: Bifurcation analysis and circuit realization for multiple-delayed Wang-Chen with hidden chaotic attractors. Nonlinear Dyn. **85**, 1635–1650 (2016)
32. Wei, Z., Wang, Z.: Chaotic behavior and modified function projective synchronization of a simple system with one stable equilibrium. Kybernetika **49**, 359–374 (2013)
33. Wei, Z., Yang, Q.: Dynamical analysis of a new autonomous 3–D chaotic system only with stable equilibria. Nonlinear Anal. Real World Appl. **12**, 106–118 (2011)
34. Wei, Z., Yang, Q.: Dynamical analysis of the generalized Sprott C system with only two stable equilibria. Nonlinear Dyn. **68**, 543–554 (2012)
35. Wei, Z.C., Pehlivan, I.: Chaos, coexisting attractors, and circuit design of the generalized Sprott C system with only two stable equilibria. Optoelectron. Adv. Mater. Rapid Commun. **6**, 742–745 (2012)
36. Westerlund, S., Ekstam, L.: Capacitor theory. IEEE Trans. Dielectr. Electr. Insul. **1**, 826–839 (1994)
37. Yalcin, M.E.: Increasing the entropy of a random number generator using n-scroll chaotic attractors. Int. J. Bifurc. Chaos **17**, 4471–4479 (2007)
38. Yang, Q., Chen, G.: A chaotic system with one saddle and two stable node-foci. Int. J. Bifurc. Chaos **18**, 1393–1414 (2008)
39. Yang, Q., Wei, Z., Chen, G.: An unusual 3D chaotic system with two stable node-foci. Int. J. Bifurc. Chaos **20**, 1061–1083 (2010)
40. Yang, Q.G., Zeng, C.B.: Chaos in fractional conjugate lorenz system and its scaling attractor. Commun. Nonlinear Sci. Numer. Simul. **15**, 4041–4051 (2010)
41. Yeniceri, R., Yalcin, M.E.: Multi-scroll chaotic attractors from a generalized time-delayed sampled-data system. Int. J. Circ. Theory Appl. **44**, 1263–1276 (2016)
42. Zidan, M.A., Radwan, A.G., Salama, K.N.: Controllable V–shape multiscroll butterfly attractors system and circuit implementation. Int. J. Bifurc. Chaos **22**, 1250,142 (2012)

Chapter 3
Systems with an Infinite Number of Equilibrium Points

3.1 Simple Systems with Line Equilibrium

We consider a general equation with quadratic nonlinearities with 12 coefficients $(a_1 - a_{12})$ as follows:

$$\begin{cases} \dot{x} = z \\ \dot{y} = a_1 x + a_2 z + a_3 z^2 + a_4 xz + a_5 yz \\ \dot{z} = a_6 x + a_7 z + a_8 x^2 + a_9 z^2 + a_{10} xy + a_{11} xz + a_{12} yz \end{cases} \quad (3.1)$$

It is noteworthy that we can find the equilibrium points of the general form (3.1) by solving

$$z = 0 \quad (3.2)$$

$$a_1 x + a_2 z + a_3 z^2 + a_4 xz + a_5 yz = 0 \quad (3.3)$$

$$a_6 x + a_7 z + a_8 x^2 + a_9 z^2 + a_{10} xy + a_{11} xz + a_{12} yz = 0 \quad (3.4)$$

By substituting Eq. (3.2) into Eqs. (3.3) and (3.4), it is easy to verify that system (3.1) has a line equilibrium

$$E(0, y, 0) \quad (3.5)$$

By applying the search routine into the general form (3.1), many systems can be found. An elegant case is the Jafari LE_8 system [3] for

$$\begin{cases} a_2 = a_3 = a_4 = a_6 = a_7 = a_9 = a_{12} = 0 \\ a_5 = -1, a_1 = a_{10} = a_{11} = 1 \\ a_8 = -a \end{cases} \quad (3.6)$$

© The Author(s) 2017
V.-T. Pham et al., *Systems with Hidden Attractors*,
SpringerBriefs in Nonlinear Circuits, DOI 10.1007/978-3-319-53721-4_3

Therefore, the Jafari LE$_8$ system is described by

$$\begin{cases} \dot{x} = z \\ \dot{y} = x - yz \\ \dot{z} = -ax^2 + xy + xz \end{cases} \tag{3.7}$$

in which a is the positive parameter. The line equilibrium of the Jafari LE$_8$ system is described as

$$E = \{(x, y, z) \in R^3 \,|\, x = 0, y = y^*, z = 0\} \tag{3.8}$$

The Jacobian matrix of the Jafari LE$_8$ system at the equilibrium E is

$$\mathbf{J}_E = \begin{bmatrix} 0 & 0 & 1 \\ 0 & 0 & -y^* \\ y^* & 0 & 0 \end{bmatrix} \tag{3.9}$$

while its characteristic equation is

$$\lambda \left(\lambda^2 - y^*\right) = 0 \tag{3.10}$$

From the characteristic equation (3.10), we obtain the following eigenvalues

$$\lambda_1 = 0, \lambda_{2,3} = \pm\sqrt{y^*} \tag{3.11}$$

The Jafari LE$_8$ generates complex dynamic for $a = 3$ as shown in Fig. 3.1.

Similarly, we consider a general equation with quadratic nonlinearities with 9 coefficients (a_1–a_9) as follows:

$$\begin{cases} \dot{x} = y \\ \dot{y} = a_1x + a_2yz \\ \dot{z} = a_3\,|x| + a_4\,|y| + a_5x + a_6y + a_7xy + a_8xz + a_9yz \end{cases} \tag{3.12}$$

Fig. 3.1 Phase portrait in the $x - y$ plane of Jafari LE$_8$ system with a *line* equilibrium for $a = 3$ and initial conditions $(x(0), y(0), z(0)) = (0, -0.3, -1)$

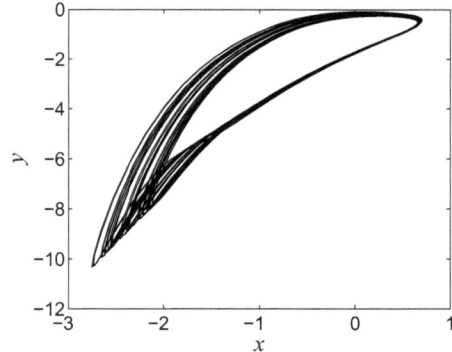

Table 3.1 Systems with a line of equilibrium points

Case	Equations	Equilibrium
A	$\dot{x} = y$	$(0, 0, z)$
	$\dot{y} = -x + yz$	
	$\dot{z} = a\,|x| - bxy - xz$	
B	$\dot{x} = y$	$(0, 0, z)$
	$\dot{y} = -ax + yz$	
	$\dot{z} = -x^2 + b\,|y| - cxy$	
C	$\dot{x} = y$	$(0, 0, z)$
	$\dot{y} = -x + yz$	
	$\dot{z} = a\,|y| - xy - bxz$	

By applying the search routine to the general form (3.12), many systems can be found. Some typical cases are reported in Table 3.1.

An elegant case is the system A in Table 3.1 for

$$\begin{cases} a_4 = a_5 = a_6 = a_9 = 0 \\ a_1 = a_8 = -1, a_2 = 1 \\ a_3 = a, a_7 = -b \end{cases} \tag{3.13}$$

Therefore, the system A is described by

$$\begin{cases} \dot{x} = y \\ \dot{y} = -x + yz \\ \dot{z} = a\,|x| - bxy - xz \end{cases} \tag{3.14}$$

in which a and b are the two positive parameters. Complex dynamic of system (3.14) is shown in Fig. 3.2.

In addition, there are few systems with lines of equilibrium in the literature (see Table 3.2). System SL_{12} has two parallel lines of equilibrium points [5] while systems

Fig. 3.2 Phase portrait in the $x - y$ plane of the system with a *line* equilibrium (3.14) for $a = 0.25, b = 19$ and initial conditions $(x(0), y(0), z(0)) = (0, -0.35, 0.45)$

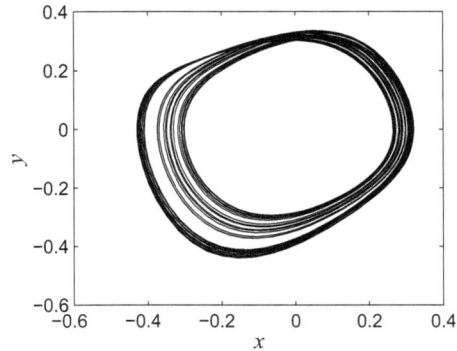

Table 3.2 Systems with lines of equilibrium points

Case	Equations	Equilibrium
SL_{12}	$\dot{x} = -y + x^2 - y^2$	$(0, 0, z)$
	$\dot{y} = -xz$	$(0, -1, z)$
	$\dot{z} = ax^2 + bxy$	$\left(\frac{ab}{a^2-b^2}, \frac{a^2}{b^2-a^2}, 0\right)$
AB_5	$\dot{x} = yz$	$\left(\pm\frac{m}{a}, \pm\frac{m}{a}, 0\right)$
	$\dot{y} = x\,\|x\| - y\,\|x\|$	$(0, y, 0)$
	$\dot{z} = m\,\|x\| - axy$	$(0, 0, z)$
AB_6	$\dot{x} = yz$	$(0, y, 0)$
	$\dot{y} = x\,\|x\| - y\,\|x\|$	$(0, 0, z)$
	$\dot{z} = a\,\|xy\| - mxy\,\|y\|$	$\left(\pm\frac{a}{m}, \pm\frac{a}{m}, 0\right)$
STR	$\dot{x} = -yz$	$(\pm 1, y, 0)$
	$\dot{y} = (ax + y + z^2)z$	$(0, 0, z)$
	$\dot{z} = x - x^3$	$\left(-1, 0, \pm\sqrt{2}\right)$

Fig. 3.3 Phase portrait in the $x - y$ plane of the system STR with *lines* of equilibrium points for $a = 2$ and initial conditions $(x(0), y(0), z(0)) = (0, 0, -0.94)$

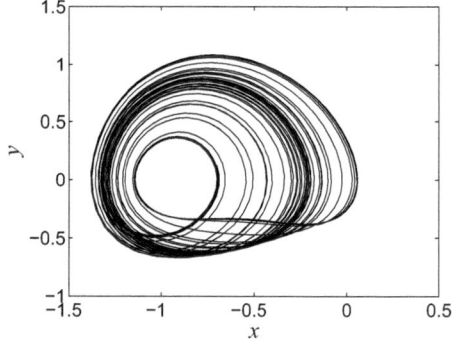

AB_5 and AB_6 have two perpendicular lines of equilibria [6]. Two systems AB_5 and AB_6 contain absolute-value functions. The amplitude of the variables in such systems can be controlled by using the coefficient m [6]. In particular, the system STR is a symmetric time-reversible flow with lines of equilibrium points [11]. Chaotic behavior of the system STR is illustrated in Fig. 3.3.

3.2 Systems with Closed Curve Equilibrium

Motivated by the discovery of systems with a line equilibrium, Gotthans and Petrzela [1] introduced a general set of three first-order differential equations to investigate the systems with circular equilibrium

$$\begin{cases} \dot{x} = az \\ \dot{y} = z f_1(x, y, z) \\ \dot{z} = x^2 + y^2 - r^2 + z f_2(x, y, z) \end{cases} \quad (3.15)$$

where a and r are free parameters, while $f_1(x, y, z)$, $f_2(x, y, z)$ are the nonlinear functions. By solving $\dot{x} = 0$, $\dot{y} = 0$, $\dot{z} = 0$, it is easy to see that the equilibrium of system (3.15) is located on a circle with a radius r

$$x^2 + y^2 = r^2 \quad (3.16)$$

By applying a search routine, authors found the following smooth functions

$$f_1(x, y, z) = bx + cz^2 \quad (3.17)$$

$$f_2(x, y, z) = dx \quad (3.18)$$

in which b, c, and d are free parameters. It is noteworthy that the system with circular equilibrium (3.15) exhibits chaotic behavior for $a = -0.1, b = 3, c = -2.2$, $d = -0.1$, and $r = 1$, as shown in Fig. 3.4.

In order to construct a new system, Kingni et al. [4] used different nonlinear functions

$$f_1(x, y, z) = 3x + z + z^2 \quad (3.19)$$

$$f_2(x, y, z) = -4yz \quad (3.20)$$

As a result, Kingni system with a circular equilibrium is given by

$$\begin{cases} \dot{x} = z \\ \dot{y} = z \left(3x + z + z^2 \right) \\ \dot{z} = x^2 + y^2 - r^2 - 4yz^2 \end{cases} \quad (3.21)$$

where r is the radius of circular equilibrium.

Fig. 3.4 Phase portrait in the $x - y$ plane of Gotthans and Petrzela system with a *circular* equilibrium for $a = -0.1, b = 3, c = -2.2$, $d = -0.1, r = 1$, and initial conditions $(x(0), y(0),$ $z(0)) = (0, 0, 0)$

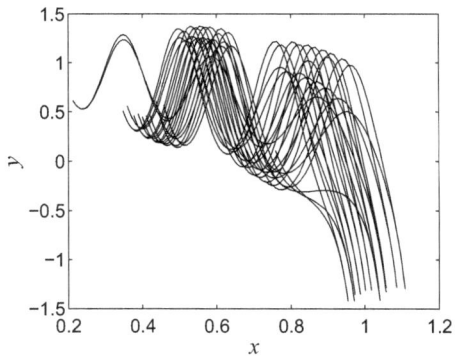

Gotthans and Petrzela proposed a general model to investigate the systems with close curve equilibrium [2]

$$\begin{cases} \dot{x} = z \\ \dot{y} = -z f_1 (x, y, z) \\ \dot{z} = f_2 (x, y) \end{cases} \tag{3.22}$$

in which $f_1(x, y, z)$ and $f_2(x, y)$ are the two nonlinear functions. Similarly, by solving $\dot{x} = 0$, $\dot{y} = 0$, and $\dot{z} = 0$, it is simple to verify that the equilibrium of system is located on a curve described by

$$f_2(x, y) = 0 \tag{3.23}$$

By choosing appropriate nonlinear functions, authors found a system with a circle of equilibrium points

$$\begin{cases} \dot{x} = z \\ \dot{y} = -z \left(ay + by^2 + xz \right) \\ \dot{z} = x^2 + y^2 - 1 \end{cases} \tag{3.24}$$

and a system with a square of equilibrium points

$$\begin{cases} \dot{x} = z \\ \dot{y} = -z \left(ay + b \left| y \right| \right) - x \left| z \right| \\ \dot{z} = |x| + |y| - 1 \end{cases} \tag{3.25}$$

Pham et al. [8, 10] studied the general model of (3.22) as follows

$$\begin{cases} \dot{x} = az \\ \dot{y} = z f_1 (x, y, z) \\ \dot{z} = f_2 (x, y) + z f_3 (x, y, z) \end{cases} \tag{3.26}$$

where a is a parameter and $f_1(x, y, z)$, $f_2(x, y)$, and $f_3(x, y, z)$ are three nonlinear functions. The equilibria of the system (3.26) are obtained by solving

$$az = 0 \tag{3.27}$$

$$z f_1(x, y, z) = 0 \tag{3.28}$$

$$f_2(x, y) + z f_3(x, y, z) = 0 \tag{3.29}$$

From (3.27), we have $z = 0$. By substituting $z = 0$ into Eqs. (3.28), (3.29), it is simple to verify that system (3.26) has an infinite number of equilibrium points in the plane $z = 0$. These equilibrium points create a curve given by

$$f_2(x, y) = 0 \tag{3.30}$$

Three nonlinear functions were selected as

$$f_1(x, y, z) = bx + cz^2 \tag{3.31}$$

$$f_2(x, y) = \left(\frac{x}{m}\right)^k + \left(\frac{y}{n}\right)^k - r^2 \tag{3.32}$$

$$f_3(x, y, z) = dx \tag{3.33}$$

in which b, c, d, k, m, n, and r are free parameters.

By changing the parameters of the function $f_3(x, y)$, various interesting shapes of equilibrium points were obtained, for example, circular, ellipse, square-shaped, rectangle-shaped and rounded-square equilibrium as illustrated in Table 3.3.

Let consider the case E

$$\begin{cases} \dot{x} = az \\ \dot{y} = z\left(bx + cz^2\right) \\ \dot{z} = x^4 + y^4 - r^2 + dxz \end{cases} \tag{3.34}$$

The equilibrium of system (3.34) is described as

$$E = \left\{ (x, y, z) \in R^3 \,\middle|\, x = x^*, y = \pm\sqrt[4]{r^2 - (x^*)^4}, z = 0 \right\} \tag{3.35}$$

The Jacobian matrix of the system at the equilibrium E is

$$\mathbf{J}_E = \begin{bmatrix} 0 & 0 & a \\ 0 & 0 & bx^* \\ 4(x^*)^3 & 4(y^*)^3 & dx^* \end{bmatrix} \tag{3.36}$$

while its characteristic equation is

$$\lambda \left(\lambda^2 - dx^*\lambda - 4bx^*\left(y^*\right)^3 - 4a\left(x^*\right)^3 \right) = 0 \tag{3.37}$$

Table 3.3 Different shapes of equilibrium points based on the nonlinear function $f_3(x, y)$

Case	$f_3(x, y)$	Equilibrium
A	$x^2 + y^2 - 1$	Circular equilibrium
B	$\left(\frac{x}{m}\right)^2 + \left(\frac{y}{n}\right)^2 - 1$	Ellipse equilibrium
C	$x^{12} + y^{12} - 1$	Square-shaped equilibrium
D	$\left(\frac{x}{m}\right)^{12} + \left(\frac{y}{n}\right)^{12} - 1$	Rectangle-shaped equilibrium
E	$x^4 + y^4 - r^2$	Rounded-square equilibrium

Fig. 3.5 Phase portrait in
the $x - y$ plane of the system
(3.34) with a *rounded-square*
equilibrium for $a = -0.1$,
$b = 3, c = -2.2, d = -0.2$,
and $r = 0.9$

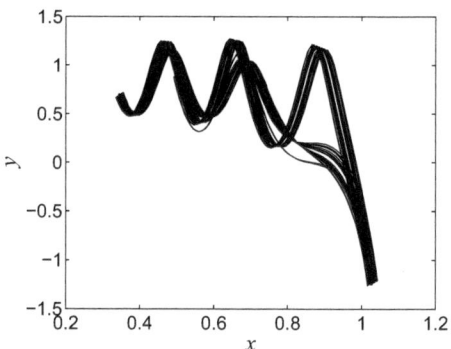

Thus, there are one zero eigenvalue and two nonzero eigenvalues which depend on
the parameters (a, b, d) and the position of the equilibrium point E.

As shown in Fig. 3.5, the system can generate chaotic behavior for $d = -0.2$

3.3 Systems with Open Curve Equilibrium

The general model (3.26) is also applied to investigate the systems with open curve
equilibrium. A system with infinite equilibria [9] was designed by selecting

$$f_1(x, y, z) = xz \tag{3.38}$$

$$f_2(x, y) = x - b|y| \tag{3.39}$$

$$f_3(x, y, z) = cy^2 - z^2 \tag{3.40}$$

where b and c are the two positive parameters. As a result, the system has the following
form

$$\begin{cases} \dot{x} = -az \\ \dot{y} = xz^2 \\ \dot{z} = x - b|y| + z\left(cy^2 - z^2\right) \end{cases} \tag{3.41}$$

where a is the positive parameter. The equilibrium points of system (3.41) are located
on a piecewise linear curve

$$E = \left\{(x, y, z) \in R^3 \mid x = b|y^*|, y = y^*, z = 0\right\} \tag{3.42}$$

Fig. 3.6 Phase portrait in the $x - y$ plane of the system (3.41) with a piecewise *linear curve* of equilibrium points for $a = 1$, $b = 0.4$, $c = 3$, and initial conditions $(x(0), y(0), z(0)) = (0.5, 0.5, 0.5)$

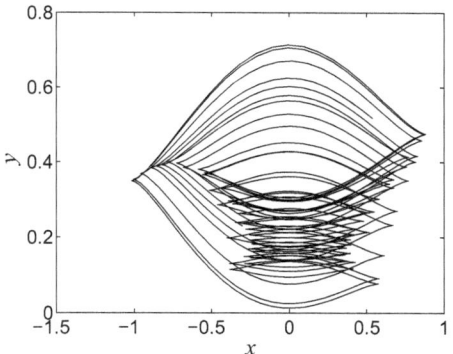

The characteristic equation of the system at the equilibrium E is

$$\lambda \left(\lambda^2 - c(y^*)^2 \lambda + a \right) = 0 \tag{3.43}$$

Thus, there are one zero eigenvalue and two nonzero eigenvalues which depend on the parameters a, c, and the position of the equilibrium. The system generates chaotic behavior for $c = 3$ (see Fig. 3.6).

A system with a parabola of equilibrium points is given as

$$\begin{cases} \dot{x} = -az \\ \dot{y} = -bz\,|z| \\ \dot{z} = x^2 + y + z\,(z - xy) \end{cases} \tag{3.44}$$

The equilibrium point $E(x^*, y^*, 0)$ of system (3.44) is located on a parabola

$$y^* = -(x^*)^2 \tag{3.45}$$

Also, a system with a hyperbola of equilibrium points is given as:

$$\begin{cases} \dot{x} = -az \\ \dot{y} = z\,(b\,|z| - 1) \\ \dot{z} = x^2 - y^2 - 1 + z\left(y^2 - z^2\right) \end{cases} \tag{3.46}$$

The equilibrium points of system (3.46) are located on a hyperbola

$$(x^*)^2 - (y^*)^2 = 1 \tag{3.47}$$

The shapes of equilibrium points (3.45) and (3.47) are illustrated in Fig. 3.7.

Fig. 3.7 Shapes of
equilibrium points of
systems (3.44), (3.46): a
parabola (*black*) and a
hyperlola (*red*), respectively

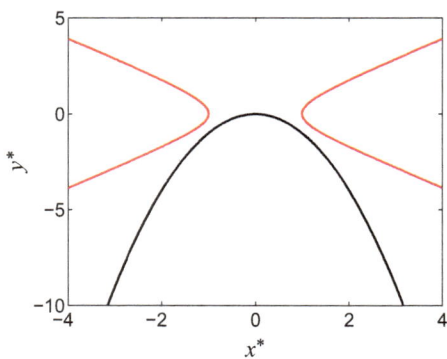

3.4   Constructing a System with Infinite Equilibria

The Molaie SE_{18} system [7] is given by

$$\begin{cases} \dot{x} = z, \\ \dot{y} = -y + z, \\ \dot{z} = -2.1x - 0.1z - y^2 + 0.11xz + 0.5yz, \end{cases} \tag{3.48}$$

The Molaie SE_{18} system has only one equilibrium $E(0, 0, 0)$. Moreover, it is a stable equilibrium because of its eigenvalues

$$\lambda_1 = -1.0000, \lambda_{2,3} = -0.0500 \pm 1.4483i \tag{3.49}$$

We select a reported memristive device described by

$$\begin{cases} \dot{w} = cy \\ h(y, w) = (bw - 1)y \end{cases} \tag{3.50}$$

in which w, y, and $h(y, w)$ are the internal state, the input, and the output of the memristive device while b and c are the two parameters of the memristive device. We add the memristive device to the system Molaie SE_{18} as follows:

$$\begin{cases} \dot{x} = z, \\ \dot{y} = -y + z + a(bw - 1)y, \\ \dot{z} = -2.1x - 0.1z - y^2 + 0.11xz + 0.5yz, \\ \dot{w} = cy, \end{cases} \tag{3.51}$$

in which the connection between the Molaie SE_{18} system and the memristive device is denoted as a. The equilibrium of system (3.51) is found by solving

$$z = 0 \tag{3.52}$$

Fig. 3.8 Phase portrait in the $x - y$ plane of the system (3.51) with an infinite number of equilibrium points for $a = 0.01, b = 0.1, c = 1$, and initial conditions $(x(0), y(0), z(0), w(0)) = (-28, 0, 0, 0)$

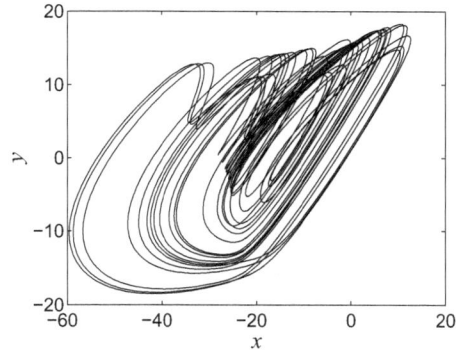

$$-y + z + a(bw - 1)y = 0 \tag{3.53}$$

$$-2.1x - 0.1z - y^2 + 0.11xz + 0.5yz = 0 \tag{3.54}$$

$$cy = 0 \tag{3.55}$$

By substituting Eqs. (3.52), (3.55) into Eqs. (3.53), (3.54), it is easy to verify that system (3.51) has an uncountable number of equilibrium points

$$E(0, 0, 0, w) \tag{3.56}$$

Interestingly, the system (3.51) exhibits complex behavior for $a = 0.01, b = 0.1$, $c = 1$, and initial conditions $(x(0), y(0), z(0), w(0)) = (-28, 0, 0, 0)$, as shown in Fig. 3.8.

3.5   Multi-scroll Attractors in a System with Infinite Equilibria

By applying a state feedback controller to a generalized autonomous Duffing–Holmes type oscillator [13], a new 4D system [12] was proposed

$$\begin{cases} \dot{x} = y \\ \dot{y} = f(x) + 0.35y - 1.95z + 0.01w \\ \dot{z} = 0.45(y - z) \\ \dot{w} = -0.1(xy + z) \end{cases} \tag{3.57}$$

in which $f(x)$ is a nonlinear function to generate multi-scroll attractors. In order to obtain an even number of scrolls, the function $f(x)$ is chosen as

Fig. 3.9 4-scroll attractor in the $x - y$ plane of the system (3.57)

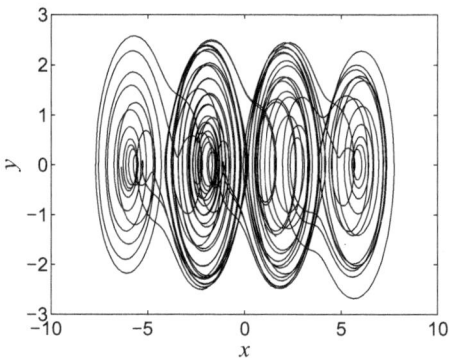

$$f(x) = f_0(x) + \sum_{n=-(N-1)}^{N-1} f_n(x) \tag{3.58}$$

where

$$\begin{cases} f_0(x) = -x \\ f_n(x) = k\tanh\left(kx + \mathrm{sgn}(n)\,k^{|n|+2}\right) \end{cases} \tag{3.59}$$

and $k = 2$ for $N \geq 1$. For generating odd number of scrolls, the function $f(x)$ is selected as

$$f(x) = f_0(x) + \sum_{n=-(N-2)}^{N-1} f_n(x) \tag{3.60}$$

where

$$\begin{cases} f_0(x) = -x - k \\ f_n(x) = k\tanh\left(kx + \mathrm{sgn}(n)\,k^{|n|+2}\right) \end{cases} \tag{3.61}$$

and $k = 2$ for $N \geq 2$.

For example, system (3.57) exhibits 4-scroll attractor as shown in Fig. 3.9.

3.6 Fractional-Order Form of Systems with Infinite Equilibria

By considering the fractional derivation effect on the system with a circular equilibrium (3.21), Kingni et al. [4] proposed its fractional-order form

$$\begin{cases} \frac{d^\alpha x}{dt^\alpha} = z \\ \frac{d^\alpha y}{dt^\alpha} = z\left(3x + z + z^2\right) \\ \frac{d^\alpha z}{dt^\alpha} = x^2 + y^2 - r^2 - 4yz^2 \end{cases} \tag{3.62}$$

Fig. 3.10 Phase portrait in the $x - y$ plane of the system fractional-order system (3.62) for $r = 0.992$, $\alpha = 0.97$ and initial conditions $(x(0), y(0), z(0)) = (-3.14, -2.2, -6.91)$

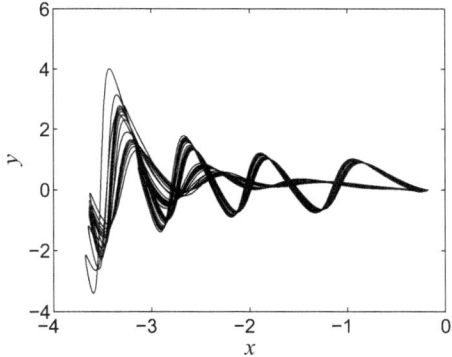

where r is the radius of circular equilibrium and α is the different fractional-order $0 < \alpha < 1$. The chaotic attractor of fractional-order system (3.62) is presented in Fig. 3.10.

Moreover, Zhou et al. introduced a 4D fractional-order system [14]

$$
\begin{cases}
\frac{d^\alpha x}{dt^\alpha} = 10\,(y - x) + w \\
\frac{d^\alpha y}{dt^\alpha} = 15x - xz \\
\frac{d^\alpha z}{dt^\alpha} = -2.5z + 4x^2 \\
\frac{d^\alpha w}{dt^\alpha} = -10y - w
\end{cases}
\tag{3.63}
$$

in which α is a fractional order. The fractional-order system (3.63) has an infinite number of equilibrium points

$$
E(0, y, 0, -10y)
\tag{3.64}
$$

The fractional-order system (3.63) can display chaotic behavior for a fractional order $\alpha = 0.95$ [14].

References

1. Gotthans, T., Petržela, J.: New class of chaotic systems with circular equilibrium. Nonlinear Dyn. **73**, 429–436 (2015)
2. Gotthans, T., Sportt, J.C., Petržela, J.: Simple chaotic flow with circle and square equilibrium. Int. J. Bifurc. Chaos **26**(1650), 137–8 (2016)
3. Jafari, S., Sprott, J.C.: Simple chaotic flows with a line equilibrium. Chaos Solitons Fract. **57**, 79–84 (2013)
4. Kingni, S.T., Pham, V.T., Jafari, S., Kol, G.R., Woafo, P.: Three-dimensional chaotic autonomous system with a circular equilibrium: analysis, circuit implementation and its fractional-order form. Circuits Syst. Signal Process. **35**(19), 331–1948 (2016)

5. Li, C., Sprott, J.C.: Chaotic flows with a single nonquadratic term. Phys. Lett. A **378**, 178–183 (2014)
6. Li, C., Sprott, J.C., Yuan, Z., Li, H.: Constructing chaotic systems with total amplitude control. Int. J. Bifurc. Chaos **25**, 1530,025–14 (2015)
7. Molaie, M., Jafari, S., Sprott, J.C., Golpayegani, S.M.R.H.: Simple chaotic flows with one stable equilibrium. Int. J. Bifurc. Chaos **23**, 1350,188 (2013)
8. Pham, V.-T., Jafari, S., Volos, C., Giakoumis, A., Vaidyanathan, S., Kapitaniak, T.: A chaotic system with equilibria located on the rounded square loop and its circuit implementation. IEEE Trans. Circuits Syst. II Express Briefs **63**, 878–882 (2016)
9. Pham, V.-T., Jafari, S., Volos, C., Vaidyanathan, S., Kapitaniak, T.: A chaotic system with infinite equilibria located on a piecewise linear curve. Optik **127**, 9111–9117 (2016)
10. Pham, V.-T., Jafari, S., Wang, X., Ma, J.: A chaotic system with different shapes of equilibria. Int. J. Bifurc. Chaos **26**, 1650,069 (2016)
11. Sprott, J.C.: Symmetric time-reversible flows with a strange attractor. Int. J. Bifurc. Chaos **25**, 1550,078–7 (2015)
12. Tahir, D.R., Jafari, S., Pham, V.T., Volos, C., Wang, X.: A novel no–equilibrium chaotic system with multiwing butterfly attractors. Int. J. Bifurc. Chaos **25**, 1550,056 (2015)
13. Tamasevicius, A., Bumeliene, S.: Autonomous Duffing-Holmes type chaotic oscillator. Electr. Electron. Eng. **93**, 43–46 (2009)
14. Zhou, P., Huang, K., Yang, C.: A fractional-order chaotic system with an infinite number of equilibrium points. Discret. Dyn. Nat. Soc. **2013**, 910,189–6 (2013)

Chapter 4
Systems Without Equilibrium

4.1 Sprott A (Nose–Hoover) System

Nose–Hoover oscillator is an important system which describes many natural phenomena [13]. Its special case is the Sprott A system [14] given by

$$\begin{cases} \dot{x} = y \\ \dot{y} = -x + yz \\ \dot{z} = -y^2 + 1 \end{cases} \tag{4.1}$$

where x, y, and z are state variables. By solving $\dot{x} = 0$, $\dot{y} = 0$, and $\dot{z} = 0$, it is simple
to verify that the Sprott A system has no equilibrium. The rate of volume expansion
of the Sprott A system is

$$\nabla V = \frac{\partial \dot{x}}{\partial x} + \frac{\partial \dot{y}}{\partial y} + \frac{\partial \dot{z}}{\partial z} = z \tag{4.2}$$

However, the Sprott A system is conservative because the sum of its Lyapunov
exponents equals zero [14]:

$$\sum_{i=1}^{3} L_i = L_1 + L_2 + L_3 = 0.0138 + 0 + (-0.0138) = 0 \tag{4.3}$$

The state plane plot for the Sprott A system is presented in Fig. 4.1.

Interestingly, the Sprott A system has variables which can be boosted [9]. It means
that we can change the DC offset of the variable to any level [7–9]. This special feature
is an vital issue in engineering applicators which requires the transformation from a
bipolar signal to a unipolar signal and vice versa [11, 16]. It is trivial to verify that
there is no effect on the dynamics of system (4.1) when replacing the variable x in
system (4.1) with the new term $x + k_x$:

© The Author(s) 2017
V.-T. Pham et al., *Systems with Hidden Attractors*,
SpringerBriefs in Nonlinear Circuits, DOI 10.1007/978-3-319-53721-4_4

Fig. 4.1 Phase portrait in
the $x - y$ plane of the Sprott
A system for the initial
conditions
$(x(0), y(0), z(0)) = (0, 5, 0)$

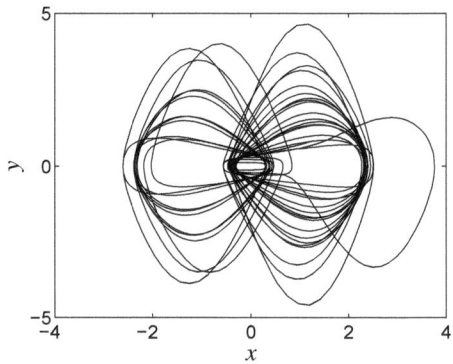

$$\begin{cases} \dot{x} = y \\ \dot{y} = -x + yz - k_x \\ \dot{z} = -y^2 + 1 \end{cases} \tag{4.4}$$

Here, k_x is a control constant. In other words, the offset of the variable x has been
provided via the control constant k_x.

By replacing the y^2 term in (4.1) with a new $|y|^a$ term, a variant of the Nose–
Hoover system was proposed

$$\begin{cases} \dot{x} = y \\ \dot{y} = -x + yz \\ \dot{z} = -|y|^a + 1 \end{cases} \tag{4.5}$$

Chaotic sea can be observed in system (4.5) for $a \geq 1$ [15].

Maaita et al. introduced a new system without equilibrium [10] based on the Sprott
A system

$$\begin{cases} \dot{x} = y \\ \dot{y} = -yz - z^3 \\ \dot{z} = y^2 - a \end{cases} \tag{4.6}$$

There is a cubic nonlinearity in Maaita system [10].

4.2 Wei System Without Equilibrium

The Sprott D system is described by

$$\begin{cases} \dot{x} = -y \\ \dot{y} = x + z \\ \dot{z} = 3y^2 + xz \end{cases} \tag{4.7}$$

Fig. 4.2 Phase portrait in the $x - y$ plane of the Wei system without equilibrium for $a = 0.35$ and initial conditions $(x(0), y(0), z(0)) = (-1.6, 0.82, 1.9)$

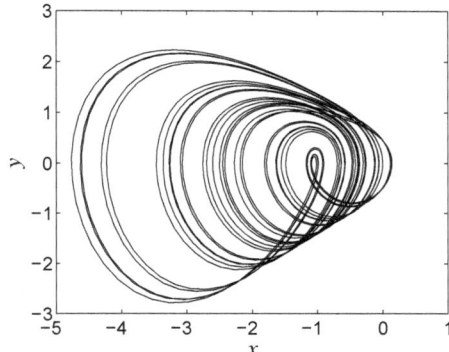

The Sprott D system has a non-hyperbolic equilibrium $E(0, 0.0, 0)$ with three eigenvalues

$$\lambda_1 = 0, \lambda_{2,3} = \pm i \tag{4.8}$$

By adding a control parameter (a) to the Sprott D, Wei system [18] was constructed as follows:

$$\begin{cases} \dot{x} = -y \\ \dot{y} = x + z \\ \dot{z} = 2y^2 + xz - a \end{cases} \tag{4.9}$$

It is simple to verify that the perturbation a changed both the type and the number of equilibria. As a result, there is no equilibrium in Wei system (4.9) for $a > 0$. Despite the absence of equilibrium, Wei system can display chaotic behavior (see Fig. 4.2).

4.3 Simple Systems with No Equilibrium

As has been discussed in Sect. 2.2, systematic search is a useful approach to find rare systems with hidden attractors, such as the ones without equilibrium [5]. In order to illustrate such interesting approach, let us take an example. We consider a general equation with quadratic nonlinearities with eight coefficients (a_1-a_8) as follows:

$$\begin{cases} \dot{x} = y \\ \dot{y} = z \\ \dot{z} = a_1y + a_2z + a_3y^2 + a_4z^2 + a_5xy + a_7xz + a_7yz + a_8 \end{cases} \tag{4.10}$$

in which

$$a_8 \neq 0 \tag{4.11}$$

It is noteworthy that we can find the equilibrium points of the general form (4.10) by solving

$$y = 0 \tag{4.12}$$

$$z = 0 \tag{4.13}$$

$$a_1 y + a_2 z + a_3 y^2 + a_4 z^2 + a_5 xy + a_7 xz + a_7 yz + a_8 = 0 \tag{4.14}$$

By substituting Eqs. (4.12) and (4.13) into Eq. (4.14), we have

$$a_8 = 0 \tag{4.15}$$

Eq. (4.15) is inconsistent; therefore, there is no equilibrium in system (4.10).

By applying the search routine to the general form (4.10) with the condition (4.11), many systems are found. One of the most elegant systems is the Jafari NE_6 system [5] for

$$\begin{cases} a_2 = a_3 = a_4 = a_5 = 0 \\ a_1 = a_6 = a_7 = -1 \\ a_8 = -0.75 \end{cases} \tag{4.16}$$

The Jafari NE_6 system is described by

$$\begin{cases} \dot{x} = y \\ \dot{y} = z \\ \dot{z} = -y - xz - yz - 0.75 \end{cases} \tag{4.17}$$

Although there is no equilibrium, the Jafari NE_6 system generates complex dynamics, as shown in Fig. 4.3.

Fig. 4.3 Phase portrait in the $x - y$ plane of the Jafari NE_6 system without equilibrium for initial conditions $(x(0), y(0), z(0) = (0, 3, -0.1)$

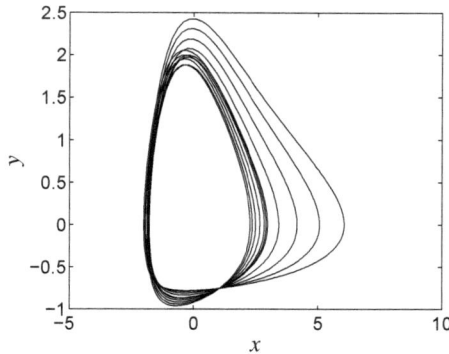

Fig. 4.4 Phase portraits in the $x - y$ plane of the system (4.18) when varying the value of the control parameter k_x: $k_x = 0$ (*black*), $k_x = -2$ (*red*)

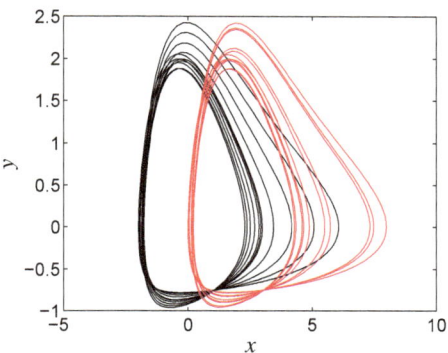

Remarkably, the state variable x appears only in the third equation of (4.17). As a result, the state variable x are controllable conveniently by replacing x with $x + k_x$. Here, k_x is a control parameter. Hence, Jafari NE$_6$ system is rewritten in the following form

$$\begin{cases} \dot{x} = y \\ \dot{y} = z \\ \dot{z} = -y - (x + k_x)\, z + yz - 0.75 \end{cases} \tag{4.18}$$

We vary the level of amplitude by changing the value of the control parameter k_x (see Fig. 4.4).

Similarly, we consider a general equation with quadratic nonlinearities with six coefficients (a_1–a_6) as follows:

$$\begin{cases} \dot{x} = y \\ \dot{y} = a_1 x + a_2 yz \\ \dot{z} = a_3 |x| + a_4 |y| + a_5 xy + a_6 \end{cases} \tag{4.19}$$

One of the most elegant systems is found for

$$\begin{cases} a_4 = 0 \\ a_1 = a_2 = -1, a_3 = a_5 = 1 \\ a_6 = -a, a > 0 \end{cases} \tag{4.20}$$

Therefore, the system [12] is described by

$$\begin{cases} \dot{x} = y \\ \dot{y} = -x - yz \\ \dot{z} = |x| + xy - a \end{cases} \tag{4.21}$$

System (4.21) displays chaotic behavior as illustrated in Fig. 4.5.

Fig. 4.5 Phase portrait in
the $x - y$ plane of the
system (4.21) for $a = 1.35$
and initial conditions
$(x(0), y(0), z(0)) =$
$(0, 0.1, 0)$

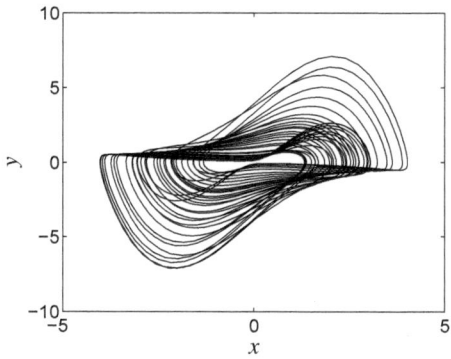

4.4 Constructing a System with No Equilibrium

The Jafari LE_5 system [6] is given by

$$\begin{cases} \dot{x} = y \\ \dot{y} = -1.5x + yz \\ \dot{z} = -x^2 + y^2 - 5xy \end{cases} \tag{4.22}$$

It is easy to see that the Jafari LE_5 system has a line of equilibria

$$E(0, 0, z) \tag{4.23}$$

It is also simple to imagine that a tiny perturbation may change the number of equilibrium points of the Jafari LE_5 system. By adding a simple parameter $a \neq 0$ to the Jafari LE_5 system, a new system is introduced as

$$\begin{cases} \dot{x} = y \\ \dot{y} = -1.5x + yz \\ \dot{z} = -x^2 + y^2 - 5xy + a \end{cases} \tag{4.24}$$

The new system possesses no equilibrium points. System without equilibrium exhibits chaos for $a = 0.001$ (see Fig. 4.6).

Similarly, a new system is obtained by adding a control parameter (a) to the Jafari LE_6 system

$$\begin{cases} \dot{x} = y \\ \dot{y} = -x + yz \\ \dot{z} = 0.04y^2 - xy - 0.1xz + a \end{cases} \tag{4.25}$$

Fig. 4.6 Phase portrait in the $x - y$ plane of the system (4.24) for $a = 0.001$ and initial conditions $(x(0), y(0), z(0)) = (0.7, 1, 0)$

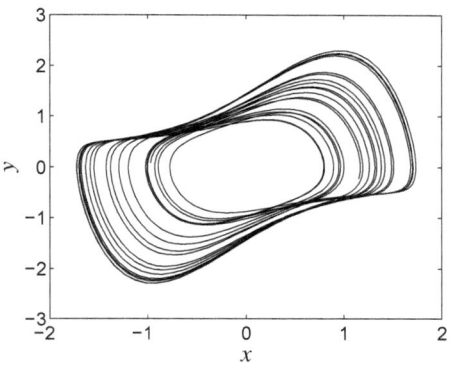

Fig. 4.7 Phase portrait in the $x - y$ plane of the system (4.25) for $a = 0.001$ and the initial conditions $(x(0), y(0), z(0)) = (1, 2, 0)$

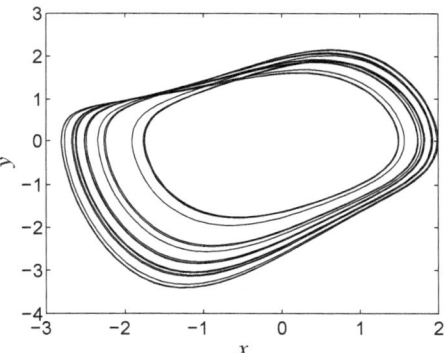

The system (4.25) has no equilibrium but can generate complex behavior as depicted in Fig. 4.7. In other words, a new hidden attractor without equilibrium is derived from another hidden attractor with infinite equilibria.

It is interesting that we can also construct a special system with different families of hidden attractors. We consider the following three dimensional system:

$$\begin{cases} \dot{x} = y \\ \dot{y} = 0.4xz - a \\ \dot{z} = 0.3y - 0.1z - 1.4y^2 - bxy - c \end{cases} \qquad (4.26)$$

in which three state variables are x, y, and z, and three positive parameters are a, b, and c. It is trivial to verify that system (4.26) is dissipative because of

$$\nabla V = \frac{\partial \dot{x}}{\partial x} + \frac{\partial \dot{y}}{\partial y} + \frac{\partial \dot{z}}{\partial z} = -0.1 < 0 \qquad (4.27)$$

The equilibrium of system (4.26) is calculated by solving $\dot{x} = 0$, $\dot{y} = 0$, $\dot{z} = 0$:

$$y = 0 \qquad (4.28)$$

$$0.4xz - a = 0 \tag{4.29}$$

$$0.3y - 0.1z - 1.4y^2 - bxy - c = 0 \tag{4.30}$$

By substituting Eq. (4.28) into Eq. (4.30), we get

$$z = -10c \tag{4.31}$$

From Eq. (4.4) we obtain:

$$xz = \frac{a}{0.4} \tag{4.32}$$

It is simple to verify that the number of equilibrium points of system depends on the values of parameters a, b, and c. Some special cases are listed as follows:

• Case A: when $a = c = 0$

Combining Eqs. (4.31) and (4.32), system (4.26) has an infinite number of equilibrium points

$$E\left(x^*, 0, 0\right) \tag{4.33}$$

• Case B: when $a \neq 0$, $c \neq 0$

From Eqs. (4.28), (4.31) and (4.32), there is only one equilibrium in system (4.26):

$$E\left(-\frac{a}{4c}, 0, -10c\right) \tag{4.34}$$

We analyze the stability of the equilibrium via its eigenvalues. The Jacobian matrix at the equilibrium point $E\left(-\frac{a}{4c}, 0, -10c\right)$ is described by

$$\mathbf{J}_E = \begin{bmatrix} 0 & 1 & 0 \\ -4c & 0 & -\frac{0.1a}{c} \\ 0 & 0.3 + \frac{ab}{4c} & -0.1 \end{bmatrix} \tag{4.35}$$

and its corresponding characteristic equation is given by

$$\lambda^3 + 0.1\lambda^2 + \left(\frac{0.12ac + 0.1a^2b}{4c^2} + 4c\right)\lambda + 0.4c = 0. \tag{4.36}$$

According to the Routh–Hurwitz stability criterion, the equilibrium E is stable because of

$$0.1\left(\frac{0.12ac + 0.1a^2b}{4c^2} + 4c\right) > 0.4c \tag{4.37}$$

In other words, system (4.26) has only one stable equilibrium in this case.

• Case C: when $a \neq 0$, $c = 0$

Fig. 4.8 Phase portrait in the $x - y$ plane of the system with variable equilibrium (4.26) for $a = 0.005, b = 0.2, c = 0$, and initial conditions $(x(0), y(0), z(0)) = (-1.53, 0.33, 0.39)$

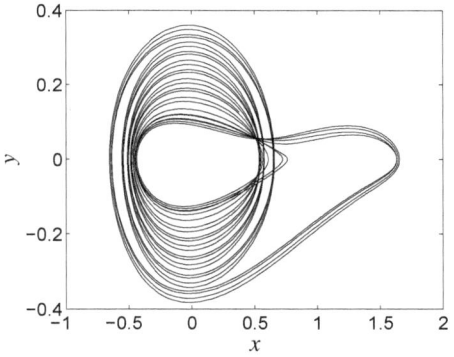

By substituting Eq. (4.31) into Eq. (4.32), we obtain

$$0 = \frac{a}{0.4} \tag{4.38}$$

Equation (4.38) is inconsistent; therefore, there is no equilibrium in system (4.26). As a result, we obtain a rare system with three main families of hidden attractors: hidden attractor with an infinite number of equilibrium points, hidden attractor with only one stable equilibrium, and hidden attractor without equilibrium. For example, the phase portrait of hidden attractor without equilibrium is illustrated in Fig. 4.8.

4.5 Multi-scroll and Multi-wing Attractors in Systems Without Equilibrium

Based on the Sprott A system, Jafari et al. constructed a novel system

$$\begin{cases} \dot{x} = y \\ \dot{y} = -x + ayz + by\sin(z) \\ \dot{z} = 1 - y^2 \end{cases} \tag{4.39}$$

where a and b are the two positive parameters [4]. System without equilibrium (4.39) generates 2×3-grid scroll chaotic sea as illustrated in Fig. 4.9, for $a = 0.1, b = 2.9$, and initial conditions $(x(0), y(0), z(0)) = (0, 5, 0)$. Interestingly, there is a hidden torus that coexists with the chaotic sea. For example, the system displays a torus for $a = 0.1, b = 2.9$, and initial conditions $(x(0), y(0), z(0)) = (0, 1, 0)$ (see Fig. 4.10).

Hu et al. proposed two systems with multi-scroll chaotic sea [3]

$$\begin{cases} \dot{x} = y \\ \dot{y} = -x + yz - a\sin(2\pi bx) \\ \dot{z} = 1 - y^2 \end{cases} \tag{4.40}$$

Fig. 4.9 2×3-grid scroll chaotic sea in the $y - z$ plane of the system (4.39) for $a = 0.1$, $b = 2.9$, and initial conditions $(x(0), y(0), z(0)) = (0, 5, 0)$

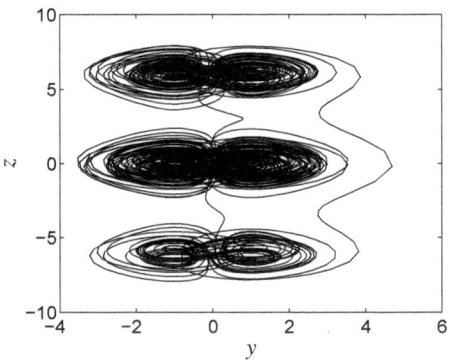

Fig. 4.10 Torus in the $x - y$ plane of the system (4.39) for $a = 0.1$, $b = 2.9$, and initial conditions $(x(0), y(0), z(0)) = (0, 1, 0)$

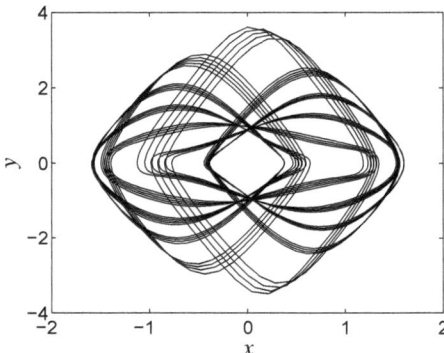

and

$$\begin{cases} \dot{x} = y \\ \dot{y} = -x + yz - 0.5a\sin{(2\pi bx)}\,(\text{sgn}\,(x - c) - \text{sgn}\,(x - d)) \\ \quad\;\; -ax\,(2 - \text{sgn}\,(x - c) + \text{sgn}\,(x - d)) \\ \dot{z} = 1 - y^2 \end{cases} \tag{4.41}$$

where a, b, c, and d are parameters.

A new 4D system [17] was designed by applying a state feedback controller to the 3D Lorenz-type system [19]

$$\begin{cases} \dot{x} = 0.55\,(y - x) \\ \dot{y} = -z\text{sgn}\,(x) + 0.1w \\ \dot{z} = f\,(x) - 1 \\ \dot{w} = -0.1y \end{cases} \tag{4.42}$$

Fig. 4.11 4-wing butterfly attractor in the $x - y$ plane of the system (4.42)

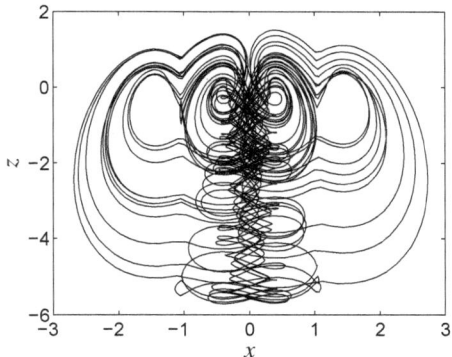

Here, $f(x)$ is a sawtooth wave function defined by

$$f(x) = f_0(x) + \sum_{n=1}^{N} f_n(x) \tag{4.43}$$

with

$$\begin{cases} f_0(x) = k|x| \\ f_n(x) = -F_n\left(2 + \text{sgn}(x - E_n) - \text{sgn}(x + E_n)\right) \\ F_n = \frac{A}{A_n}, \quad E_n = \frac{nA}{k} \end{cases} \tag{4.44}$$

for $1 \leq n \leq N$.

System (4.42) generates $2N + 2$ wings in a butterfly attractor. For example, system (4.42) exhibits 4-wing butterfly attractor for $N = 1, A = 2.6, k = 2.5, A_1 = 2, F_1 = 1.3$, and $E_1 = 1.04$ (see Fig. 4.11).

4.6 Fractional-Order Form of Systems Without Equilibrium

Starting from the integer-order counterpart (4.17), Cafagna and Grassi [1, 2] has derived the fractional-order system with no equilibrium points

$$\begin{cases} \frac{d^{\alpha_1}x}{dt^{\alpha_1}} = y \\ \frac{d^{\alpha_2}x}{dt^{\alpha_2}} = z \\ \frac{d^{\alpha_3}x}{dt^{\alpha_3}} = -y - xz - yz - 0.75 \end{cases} \tag{4.45}$$

Fig. 4.12 Chaotic attractor in the $x - y$ plane of the fractional-order system without equilibrium (4.45) for commensurate fractional orders $\alpha_1 = \alpha_2 = \alpha_3 = \alpha = 0.98$ and initial conditions $(x(0), y(0), z(0)) = (0, 3, -0.1)$

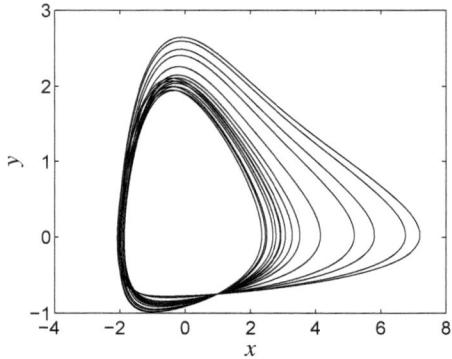

in which α_1, α_2, and α_3 are the derivative orders ($0 < \alpha_1, \alpha_2, \alpha_3 < 1$). Figure 4.12 shows the phase portrait of the fractional-order system (4.45) with commensurate fractional orders ($\alpha_1 = \alpha_2 = \alpha_3 = \alpha = 0.98$).

References

1. Cafagna, D., Grassi, G.: Elegant chaos in fractional–order system without equilibria. Math. Prob. Eng. 380,436 (2013)
2. Cafagna, D., Grassi, G.: Chaos in a fractional-order system without equilibrium points. Commun. Nonlinear Sci. Numer. Simul. **19**, 2919–2927 (2014)
3. Hu, X., Liu, C., Liu, L., Ni, J., Li, S.: Multi-scroll hidden attractors in impoved Sprott A system. Nonlinear Dyn. **86**, 1725–1734 (2016)
4. Jafari, S., Pham, V.T., Kapitaniak, T.: Multiscroll chaotic sea obtained from a simple 3D system without equilibrium. Int. J. Bifurcat. Chaos **26**, 1650,031 (2016)
5. Jafari, S., Sprott, J., Golpayegani, S.M.R.H.: Elementary quadratic chaotic flows with no equilibria. Phys. Lett. A **377**, 699–702 (2013)
6. Jafari, S., Sprott, J.C.: Simple chaotic flows with a line equilibrium. Chaos Solitons Fract. **57**, 79–84 (2013)
7. Li, C., Sprott, J.C.: Amplitude control approach for chaotic signals. Nonlinear Dyn. **73**, 1335–1341 (2013)
8. Li, C., Sprott, J.C.: Finding coexisting attractors using amplitude control. Nonlinear Dyn. **78**, 2059–2064 (2014)
9. Li, C., Sprott, J.C.: Variable-boostable chaotic flows. Optik **127**, 10389–10398 (2016)
10. Maaita, J.O., Volos, C.K., Kyprianidis, I.M., Stouboulos, I.N.: The dynamics of a cubic nonlinear system with no equilibrium point. J. Nonlinear Dyn. **2015**, 257,923 (2015)
11. Obeid, I., Morizio, J.C., Moxon, K.A., Nicolelis, M.A.L., Wolf, P.D.: Two multichannel integrated circuits for neural recording and signal processing. IEEE Trans. Biomed. Eng. **50**, 255–258 (2003)
12. Pham, V.T., Volos, C., Jafari, S., Kapitaniak, T.: Coexistence of hidden chaotic attractors in a novel no–equilibrium system. Nonlinear Dyn. 1–10 (2016). doi:10.1007/s11071-016-3170-x
13. Posch, H.A., Hoover, W.G., Vesely, F.J.: Canonical dynamics of the nose oscillator: stability, order, and chaos. Phys. Rev. A **33**, 4253–4265 (1986)
14. Sprott, J.: Some simple chaotic flows. Phys. Rev. E **50**, R647–650 (1994)
15. Sprott, J.C.: Elegant Chaos Algebraically Simple Chaotic Flows. World Scientific, Singapore (2010)

16. Steinhaus, B.M.: Estimating cardiac transmembrane activation and recovery times from unipolar and bipolar extracellular electrograms: a simulation study. Circ. Res. **64**, 449–462 (1989)
17. Tahir, D.R., Jafari, S., Pham, V.T., Volos, C., Wang, X.: A novel no–equilibrium chaotic system with multiwing butterfly attractors. Int. J. Bifurcat. Chaos **25**, 1550,056 (2015)
18. Wei, Z.: Dynamical behaviors of a chaotic system with no equilibria. Phys. Lett. A **376**, 102–108 (2011)
19. Yu, S., Tang, W.K.S., Lu, J., Chen, G.: Design and implementation of multi-wing butterfly chaotic attractors via Lorenz-type systems. Int. J. Bifurcat. Chaos **20**, 29–41 (2010)

Chapter 5
Synchronization of Systems with Hidden Attractors

5.1 Synchronization via Diffusion Coupling

As it is mentioned in the first chapter, one of the simplest synchronization schemes is the synchronization via diffusion coupling [11, 18, 30]. Moreover, diffusive coupling is the most common type of coupling appearing in real systems. For example, the synchronization of neuronal activity is affected by potassium lateral diffusion coupling [19]. It is interesting that various types of synchronization appear in systems with diffusion coupling. For instance, complete synchronization, inverse lag synchronization, and inverse π–lag synchronization were observed in bidirectionally coupled double-scroll circuits [29]. The route from synchronization to desynchronization of two identical bidirectionally coupled circuits was examined experimentally in [17]. Volos et al. investigated the anti-phase and inverse π–lag synchronization in coupled Duffing-type circuits [28]. Ray et al. introduced the generation of amplitude death and rhythmogenesis in coupled hidden attractor systems [24]. In addition, reaction-diffusion systems have the ability to generate various complex phenomena such as autowaves, labyrinths, and Turing pattern formation. [2, 6, 7, 9, 10, 21–23]. Thus, diffusion coupling of systems with hidden attractors is discussed in this section.

5.1.1 Diffusion Coupling of Two Systems with One Stable Equilibrium

We consider two identical nonlinear systems with only one stable equilibrium (2.2) which are linearly coupled. In other words, we have implemented bidirectional or mutual couplings as follows:

© The Author(s) 2017
V.-T. Pham et al., *Systems with Hidden Attractors*,
SpringerBriefs in Nonlinear Circuits, DOI 10.1007/978-3-319-53721-4_5

Fig. 5.1 Bifurcation
diagram of $x_2 - x_1$ when
changing the value of the
coupling coefficient ξ
between the two systems
with one stable equilibrium

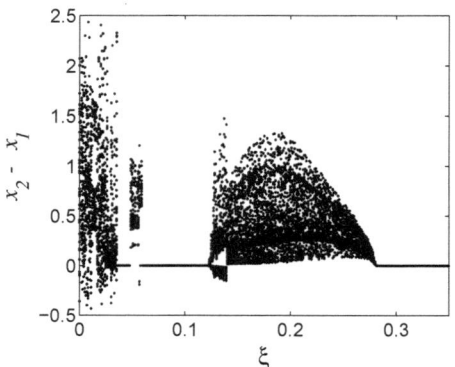

$$\begin{cases} \dot{x}_1 = y_1 z_1 + a + \xi_x \left(x_2 - x_1 \right), \\ \dot{y}_1 = x_1^2 - y_1 + \xi_y \left(y_2 - y_1 \right), \\ \dot{z}_1 = 1 - 4x_1 + \xi_z \left(z_2 - z_1 \right), \\ \dot{x}_2 = y_2 z_2 + a + \xi_x \left(x_1 - x_2 \right), \\ \dot{y}_2 = x_2^2 - y_2 + \xi_y \left(y_1 - y_2 \right), \\ \dot{z}_2 = 1 - 4x_2 + \xi_z \left(z_1 - z_2 \right), \end{cases} \tag{5.1}$$

in which ξ_x, ξ_y, and ξ_z are the coupling coefficients.

In this work, we select

$$\begin{cases} \xi_x = 0, \\ \xi_y = \xi_z = \xi. \end{cases} \tag{5.2}$$

As a result, the effect of the coupling coefficient ξ is discovered via the bifurcation diagram as shown in Fig. 5.1. As can be seen in Fig. 5.1, the system is in a desynchronization mode for small values of the coupling coefficient. Synchronization is observed when increasing the coupling coefficient ξ. There is the presence of the desynchronization–synchronization transition as well as the synchronization–desynchronization transition.

5.1.2   Diffusion Coupling of Two Systems with Infinite Equilibria

We consider two identical nonlinear systems with infinite equilibria (3.41) which are linearly coupled. It means that we have implemented bidirectional or mutual couplings in the following form:

Fig. 5.2 Bifurcation diagram of $x_2 - x_1$ when changing the value of the coupling coefficient ξ between the two systems with infinite equilibria

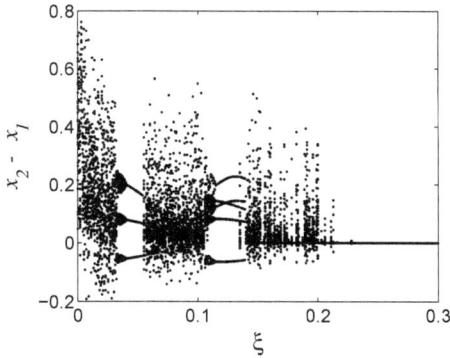

$$\begin{cases} \dot{x}_1 = -az_1 + \xi_x (x_2 - x_1), \\ \dot{y}_1 = x_1 z_1^2 + \xi_y (y_2 - y_1), \\ \dot{z}_1 = x_1 - b |y_1| + z_1 \left(cy_1^2 - z_1^2\right) + \xi_z (z_2 - z_1), \\ \dot{x}_2 = -az_2 + \xi_x (x_1 - x_2), \\ \dot{y}_2 = x_2 z_2^2 + \xi_y (y_1 - y_2), \\ \dot{z}_2 = x_2 - b |y_2| + z_2 \left(cy_2^2 - z_2^2\right) + \xi_z (z_1 - z_2), \end{cases} \tag{5.3}$$

in which ξ_x, ξ_y and ξ_z are the coupling coefficients.

In this work, we select

$$\begin{cases} \xi_y = 0, \\ \xi_x = \xi_z = \xi. \end{cases} \tag{5.4}$$

As a result, the effect of the coupling coefficient ξ is discovered through the bifurcation diagram. The bifurcation diagram is presented in Fig. 5.2 when increasing the value of the coupling coefficient ξ from 0 to 0.3. As can be seen from Fig. 5.2, the system is in a desynchronization mode for small values of the coupling coefficient. Different regions, in which coupling system shows chaotic, periodic behavior, appear for $\xi < 0.2$. There is a transition from desynchronization to synchronization when increasing the coupling coefficient ξ. Finally, the complete synchronization is observed.

5.1.3 Diffusion Coupling of Two Systems Without Equilibrium

We consider two identical nonlinear systems without equilibrium (4.21) which are linearly coupled. In other words, we have implemented bidirectional or mutual couplings as follows:

Fig. 5.3 Bifurcation
diagram of $x_2 - x_1$ when
changing the value of the
coupling coefficient ξ
between the two systems
without equilibrium

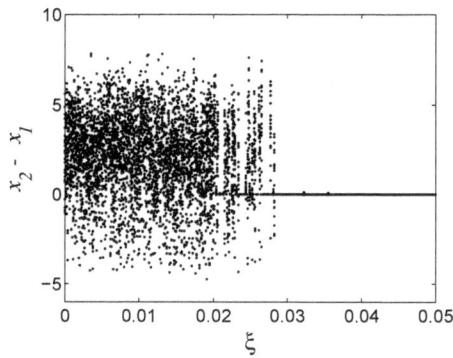

$$\begin{cases} \dot{x}_1 = y_1 + \xi_x \left(x_2 - x_1 \right), \\ \dot{y}_1 = -x_1 - y_1 z_1 + \xi_y \left(y_2 - y_1 \right), \\ \dot{z}_1 = |x_1| + x_1 y_1 - a + \xi_z \left(z_2 - z_1 \right), \\ \dot{x}_2 = y_2 + \xi_x \left(x_1 - x_2 \right), \\ \dot{y}_2 = -x_2 - y_2 z_2 + \xi_y \left(y_1 - y_2 \right), \\ \dot{z}_2 = |x_2| + x_2 y_2 - a + \xi_z \left(z_1 - z_2 \right), \end{cases} \tag{5.5}$$

in which ξ_x, ξ_y, and ξ_z are the coupling coefficients.

In this work, we select

$$\begin{cases} \xi_x = 0, \\ \xi_y = \xi_z = \xi. \end{cases} \tag{5.6}$$

Similarly, by changing the values of ξ in the range $[0, 0.05]$, we can investigate the effect of the coupling coefficient ξ. As a result, the bifurcation diagram is shown in Fig. 5.3. As can be seen in Fig. 5.3, the system is in a desynchronization mode for small values of the coupling coefficient ($\xi < 0.02$). In the region $0.02 \leq \xi \leq 0.03$, the system can be in either a chaotic desynchronization state or in a chaotic synchronization state, depending on the value of the coupling coefficient. However, complete synchronization is observed for $\xi > 0.03$.

5.2 Synchronization via Nonlinear Control

From the control point of view, synchronization issue deals with the synchronization of a pair of systems named the master and the slave systems [4, 14]. The aim of published works is to design control laws which guarantee that the output of the slave system tracks the output of the master system [35].

Various methodologies have been investigated for the synchronization of systems such as active control [16, 27], passive control [31], sampled-data feedback control [25, 32], sliding mode control [1, 34], backstepping control [12, 33], Takagi–Sugeno fuzzy control [26], or adaptive control [8, 13]. Furthermore, synchronization via control technique has been applied in different practical applications, especially secure communication in which direct coupling cannot be done [3, 5, 20]. In this section, the synchronization of the systems with hidden attractors is illustrated through designing nonlinear control.

5.2.1 Synchronization of Systems with One Stable Equilibrium

In this section, we discover the synchronization of two systems with stable equilibria (2.2), called the master system and the slave system, by using an adaptive controller.

We consider the following master system with the unknown system parameters a, b:

$$\begin{cases} \dot{x}_1 = y_1 z_1 + a \\ \dot{y}_1 = x_1^2 - y_1 \\ \dot{z}_1 = 1 - b x_1 \end{cases} \tag{5.7}$$

The slave system with adaptive control $\mathbf{u} = \begin{bmatrix} u_x, u_y, u_z \end{bmatrix}^T$ is given as:

$$\begin{cases} \dot{x}_2 = y_2 z_2 + a + u_x \\ \dot{y}_2 = x_2^2 - y_2 + u_y \\ \dot{z}_2 = 1 - b x_2 + u_z \end{cases} \tag{5.8}$$

The state errors between the slave system and the master system are calculated by

$$\begin{cases} e_x = x_2 - x_1 \\ e_y = y_2 - y_1 \\ e_z = z_2 - z_1 \end{cases} \tag{5.9}$$

The parameter estimation error is defined as follows:

$$e_b = b - \hat{b} \tag{5.10}$$

in which $\hat{b}$ is the estimation of the unknown parameter b.

In order to synchronize the slave system and the master system, the adaptive control is constructed in the following form:

$$\begin{cases} u_x = y_1 z_1 - y_2 z_2 - k_x e_x \\ u_y = x_1^2 - x_2^2 + e_y - k_y e_y \\ u_z = \hat{b} e_x - k_z e_z \end{cases} \tag{5.11}$$

in which k_x, k_y, and k_z are three positive gain constants and the parameter update law is described by

$$\dot{b} = -e_x \tag{5.12}$$

By applying Lyapunov stability theory, we will prove that the master system (5.7) and the slave system (5.8) are synchronized when using the adaptive control (5.11).

In this work, the Lyapunov function is selected as

$$V\left(e_x, e_y, e_z, e_b\right) = \frac{1}{2}\left(e_x^2 + e_y^2 + e_z^2 + e_b^2\right) \tag{5.13}$$

Therefore, the differentiation of V is

$$\dot{V} = e_x \dot{e}_x + e_y \dot{e}_y + e_z \dot{e}_z + e_b \dot{e}_b \tag{5.14}$$

From (5.9) and (5.10), we have

$$\begin{cases} \dot{e}_x = -k_x e_x \\ \dot{e}_y = -k_y e_y \\ \dot{e}_z = -e_b e_x - k_z e_z \end{cases} \tag{5.15}$$

and

$$\dot{e}_b = -\dot{b} \tag{5.16}$$

By substituting (5.15) and (5.16) in to (5.14), the differentiation of V can be expressed as

$$\dot{V} = -k_x e_x^2 - k_y e_y^2 - k_z e_z^2 \tag{5.17}$$

Because $\dot{V}$ is a negative semi-definite function, it is simply verified that $e_x \to 0$, $e_y \to 0$, and $e_z \to 0$ exponentially as $t \to \infty$ according to Barbalat's lemma [15]. In other words, we obtain the complete synchronization between the master system and the slave system.

We take an example to illustrate the calculation of the synchronization scheme. The parameter values of the master system and slave system are fixed as

$$a = 0.006, \quad b = 4 \tag{5.18}$$

We assume that the initial states of the master system are

$$x_1(0) = 1, \quad y_1(0) = 1, \quad z_1(0) = 1 \tag{5.19}$$

while the initial states of the slave system are

$$x_2(0) = 1.5, \quad y_2(0) = 0.5, \quad z_2(0) = 0.1 \tag{5.20}$$

The positive gain constants are chosen as follows: $k_x = 4$, $k_y = 4$, and $k_z = 4$. We take the initial condition of the parameter estimate as

Fig. 5.4 Time-history of the synchronization errors which indicates the synchronization between the master system with stable equilibrium and the slave system with stable equilibrium

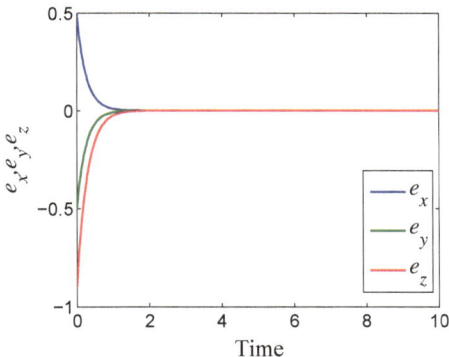

$$\hat{b}(0) = 3 \tag{5.21}$$

The time-history of the synchronization errors e_x, e_y, e_z is shown in Fig. 5.4. It is straightforward to verify that Fig. 5.4 depicts the synchronization of the master and slave systems.

5.2.2 Synchronization of Systems with Infinite Equilibrium

In this section, we discover the synchronization of two systems with infinite equilibria (3.41), called the master system and the slave system, by using an adaptive controller.

We consider the following master system with the unknown system parameters a, b and c

$$\begin{cases} \dot{x} = -az_1 \\ \dot{y} = x_1 z_1^2 \\ \dot{z} = x_1 - b\,|y_1| + cy_1^2 z_1 - z_1^3 \end{cases} \tag{5.22}$$

The slave system is given as:

$$\begin{cases} \dot{x}_2 = -az_2 + u_x \\ \dot{y}_2 = x_2 z_2^2 + u_y \\ \dot{z}_2 = x_2 - b\,|y_2| + cy_2^2 z_2 - z_2^3 + u_z \end{cases} \tag{5.23}$$

where $\mathbf{u} = \left[u_x, u_y, u_z\right]^T$ is the adaptive control. The state errors between the slave system and the master system are calculated by

$$\begin{cases} e_x = x_2 - x_1 \\ e_y = y_2 - y_1 \\ e_z = z_2 - z_1 \end{cases} \tag{5.24}$$

The parameter estimation error is defined as follows:

$$\begin{cases} e_a = a - \hat{a} \\ e_b = b - \hat{b} \\ e_c = c - \hat{c} \end{cases} \tag{5.25}$$

in which $\hat{a}$, $\hat{b}$, and $\hat{c}$ are the estimations of the unknown parameter a, b, and c.

In order to synchronize the slave system and the master system, the adaptive control is constructed in the following form:

$$\begin{cases} u_x = \hat{a}e_z - k_x e_x \\ u_y = -x_2 z_2^2 + x_1 z_1^2 - k_y e_y \\ u_z = -e_x + \hat{b}\left(|y_2| - |y_1|\right) - \hat{c}\left(y_2^2 z_2 - y_1^2 z_1\right) - z_1^3 + z_2^3 - k_z e_z \end{cases} \tag{5.26}$$

in which k_x, k_y, and k_z are three positive gain constants and the parameter update law is described by

$$\begin{cases} \dot{\hat{a}} = -e_x e_z \\ \dot{\hat{b}} = \left(|y_1| - |y_2|\right) e_z \\ \dot{\hat{c}} = \left(y_2^2 z_2 - y_1^2 z_1\right) e_z \end{cases} \tag{5.27}$$

By applying Lyapunov stability theory, we will prove that the master system (5.22) and the slave system (5.23) are synchronized when using the adaptive control (5.26).

In this work, the Lyapunov function is selected as

$$V\left(e_x, e_y, e_z, e_a, e_b, e_c\right) = \frac{1}{2}\left(e_x^2 + e_y^2 + e_z^2 + e_a^2 + e_b^2 + e_c^2\right) \tag{5.28}$$

Therefore, the differentiation of V is

$$\dot{V} = e_x \dot{e}_x + e_y \dot{e}_y + e_z \dot{e}_z + e_a \dot{e}_a + e_b \dot{e}_b + e_c \dot{e}_c \tag{5.29}$$

From (5.24) and (5.25), we have

$$\begin{cases} \dot{e}_x = -e_a e_z - k_x e_x \\ \dot{e}_y = -k_y e_y \\ \dot{e}_z = -e_b\left(|y_2| - |y_1|\right) + e_c\left(y_2^2 z_2 - y_1^2 z_1\right) - k_z e_z \end{cases} \tag{5.30}$$

and

$$\begin{cases} \dot{e}_a = -\dot{\hat{a}} \\ \dot{e}_b = -\dot{\hat{b}} \\ \dot{e}_c = -\dot{\hat{c}} \end{cases} \tag{5.31}$$

By substituting (5.30) and (5.31) in to (5.29), the differentiation of V can be expressed as

$$\dot{V} = -k_x e_x^2 - k_y e_y^2 - k_z e_z^2 \tag{5.32}$$

Because $\dot{V}$ is a negative semi-definite function, it is simply verified that $e_x \to 0$, $e_y \to 0$, and $e_z \to 0$ exponentially as $t \to \infty$ according to Barbalat's lemma [15]. In other words, we obtain the complete synchronization between the master system and the slave system.

We take an example to illustrate the calculation of the synchronization scheme. The parameter values of the master system and slave system are fixed as

$$a = 1, \quad b = 0.4, \quad c = 3 \tag{5.33}$$

We assume that the initial states of the master system is

$$x_1(0) = 0.5, \quad y_1(0) = 0.5, \quad z_1(0) = 0.5 \tag{5.34}$$

while the initial states of the slave system is

$$x_2(0) = 1, \quad y_2(0) = 0.7, \quad z_2(0) = 0.1 \tag{5.35}$$

The positive gain constants are chosen as follows: $k_x = 4$, $k_y = 4$, and $k_z = 4$. We take the initial condition of the parameter estimate as

$$\hat{a}(0) = 0.5, \quad \hat{b}(0) = 0.2, \quad \hat{c}(0) = 2.5 \tag{5.36}$$

The time-history of the synchronization errors e_x, e_y, e_z is shown in Fig. 5.5. It is straightforward to verify that Fig. 5.5 depicts the synchronization of the master and the slave systems.

Fig. 5.5 Time-history of the synchronization errors which indicates the synchronization between the master system with open curve equilibrium and the slave system with open curve equilibrium

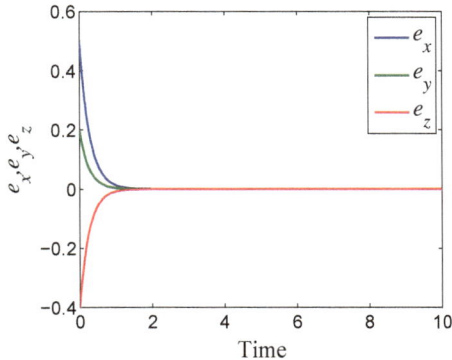

5.2.3 Synchronization of Systems Without Equilibrium

In this section, we discover the anti-synchronization of two systems without equilibrium (4.21), called the master system and the slave system, by using an adaptive controller.

We consider the following master system with the unknown system parameter a

$$\begin{cases} \dot{x}_1 = y_1 \\ \dot{y}_1 = -x_1 - y_1 z_1 \\ \dot{z}_1 = |x_1| + x_1 y_1 - a \end{cases} \tag{5.37}$$

The slave system with adaptive control $\mathbf{u} = \begin{bmatrix} u_x, u_y, u_z \end{bmatrix}^T$ is given as:

$$\begin{cases} \dot{x}_2 = y_2 + u_x \\ \dot{y}_2 = -x_2 - y_2 z_2 + u_y \\ \dot{z}_2 = |x_2| + x_2 y_2 - a + u_z \end{cases} \tag{5.38}$$

The state errors between the slave system and the master system are calculated by

$$\begin{cases} e_x = x_2 + x_1 \\ e_y = y_2 + y_1 \\ e_z = z_2 + z_1 \end{cases} \tag{5.39}$$

The parameter estimation error is defined as follows:

$$e_a = a - \hat{a} \tag{5.40}$$

in which $\hat{a}$ is the estimation of the unknown parameter a.

In order to anti-synchronize the slave system and the master system, the adaptive control is constructed in the following form:

$$\begin{cases} u_x = -e_y - k_x e_x \\ u_y = e_x + y_1 z_1 + y_2 z_2 - k_y e_y \\ u_z = -|x_1| - |x_2| - x_1 y_1 - x_2 y_2 + 2\hat{a} - k_z e_z \end{cases} \tag{5.41}$$

in which k_x, k_y, and k_z are three positive gain constants, and the parameter update law is described by

$$\dot{\hat{a}} = -2e_z \tag{5.42}$$

By applying Lyapunov stability theory, we will prove that the master system (5.37) and the slave system (5.38) are anti-synchronized when using the adaptive control (5.41).

In this work, the Lyapunov function is selected as

$$V\left(e_x, e_y, e_z, e_a\right) = \frac{1}{2}\left(e_x^2 + e_y^2 + e_z^2 + e_a^2\right) \tag{5.43}$$

Therefore, the differentiation of V is

$$\dot{V} = e_x\dot{e}_x + e_y\dot{e}_y + e_z\dot{e}_z + e_a\dot{e}_a \tag{5.44}$$

From (5.39) and (5.40), we have

$$\begin{cases} \dot{e}_x = -k_x e_x \\ \dot{e}_y = -k_y e_y \\ \dot{e}_z = -2e_a - k_z e_z \end{cases} \tag{5.45}$$

and

$$\dot{e}_a = -\dot{\hat{a}} \tag{5.46}$$

By substituting (5.45) and (5.46) in to (5.44), the differentiation of V can be expressed as

$$\dot{V} = -k_x e_x^2 - k_y e_y^2 - k_z e_z^2 \tag{5.47}$$

Because $\dot{V}$ is a negative semi-definite function, it is simply verified that $e_x \to 0$, $e_y \to 0$, and $e_z \to 0$ exponentially as $t \to \infty$ according to Barbalat's lemma [15]. In other words, we obtain the anti-synchronization between the master system and the slave system.

We take an example to illustrate the calculation of the anti-synchronization scheme. The parameter values of the master system and slave system are fixed as follows:

$$a = 1.35 \tag{5.48}$$

We assume that the initial states of the master system are

$$x_1(0) = 0, \quad y_1(0) = 0.1, \quad z_1(0) = 0 \tag{5.49}$$

while the initial states of the slave system are

$$x_2(0) = 1, \quad y_2(0) = 0.4, \quad z_2(0) = -0.5 \tag{5.50}$$

The positive gain constants are chosen as follows: $k_x = 4$, $k_y = 4$, and $k_z = 4$. We take the initial condition of the parameter estimate as

$$\hat{a}(0) = 1.3 \tag{5.51}$$

Fig. 5.6 Time-history of the synchronization errors which indicates the anti-synchronization between the master system without equilibrium and the slave system without equilibrium

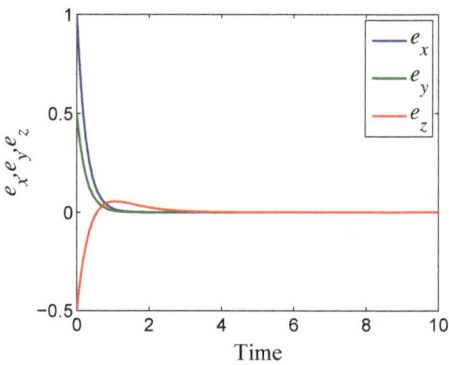

The time-history of the synchronization errors e_x, e_y, e_z is shown in Fig. 5.6. It is straightforward to verify that Fig. 5.6 depicts the anti-synchronization of the master and the slave systems.

References

1. Ablay, G.: Sliding mode control of uncertain unified chaotic systems. Nonlinear Anal. Hybrid Sys. **3**, 531–535 (2009)
2. Arena, P., Caponetto, R., Fortuna, L., Manganaro, G.: Cellular neural networks to explore complexity. Soft Comput. **1**, 120–236 (1997)
3. Banerjee, S.: Chaos Synchronization and Cryptography for Secure Communication. IGI Global, USA (2010)
4. Boccaletti, S., Kurths, J., Osipov, G., Valladares, D., Zhou, C.: The synchronization of chaotic system. Phys. Rep. **366**, 1–101 (2002)
5. Chang, J.F., Liao, T.L., Yan, J.J., Chen, H.C.: Implementation of synchronized chaotic Lu systems and its application in secure communication using PSO-based PI controller. Circuits Syst. Signal Process. **29**, 527–538 (2010)
6. Chua, L.O., Hasler, M., Moschytz, G.S., Neirynck, J.: Autonomous cellular neural networks: a unified paradigm for pattern formation and active wave propagation. IEEE Trans. Circuits Syst.–I: Fund. Th. Appl. **42**, 559–577 (1995)
7. Chua, L.O., Roska, T.: Cellular Neural Networks and Visual Computing. Cambridge University Press, Cambridge (2002)
8. Feng, G., Chen, G.: Adaptive control of discrete-time chaotic systems: a fuzzy control approach. Chaos Solitons Fractals **23**, 459–467 (2005)
9. Fortuna, L., Arena, P., Balya, D., Zarandy, A.: Cellular neural networks: a paradigm for nonlinear spatio-temporal processing. Circuits Syst. Mag. **1**, 6–21 (2001)
10. Goras, L., Chua, L.O., Leenaerts, D.M.W.: Turing patterns in CNNs—Part i: Once over lightly. IEEE Trans. Circuits Syst.–I: Fund. Th. Appl. **42**, 602–611 (1995)
11. Guemez, J., Matias, M.A.: Modified method for synchronizing and cascading chaotic system. Phys. Rev. E **52**, R2145–R2148 (1995)
12. Harb, A.M., Zaher, A.A., Al-Qaisia, A.A., Zohdy, M.A.: Recursive backstepping control of chaotic Duffing oscillators. Chaos Solitons Fractals **34**, 639–645 (2007)
13. Hua, C., Guan, X.: Adaptive control for chaotic systems. Chaos Solitons Fractals **22**, 55–60 (2004)

14. Jovic, B.: Synchronization Techniques for Chaotic Communication Systems. Springer, Germany (2011)
15. Khalil, H.K.: Nonlinear Systems, 3rd edn. Prentice Hall, New Jersey, USA (2002)
16. Kocamaz, U.E., Uyaroglu, Y.: Synchronization of vilnious chaotic oscillators with active and passive control. J. Circuit Syst. Comp. **23**, 1450,103 (2014)
17. Kyprianidis, I., Volos, C.K., Stavrinides, S.G., Anagnostopoulos, A.N.: On-off intermittent synchronization between two bidirectionally coupled double scroll circuits. Commun. Nonlinear Sci. Numer. Simul. **15**, 2192–2200 (2010)
18. Mosekilde, E., Postnov, D., Maistrenko, Y.: Chaotic Synchronization: Applications to Living Systems. World Scientific, Singapore (2002)
19. Park, E.H., Feng, Z., Durand, D.M.: Diffusive coupling and network periodicity: a computational study. Biophys. J. **95**, 1126–1137 (2008)
20. Pecora, L., Carroll, T.L.: Synchronization in chaotic systems. Phys. Rev. Lett. **64**, 821–824 (1990)
21. Perez-Munuzuri, A., Perez-Munuzuri, V., Perez-Villar, V., Chua, L.O.: Spiral waves on a 2–D array of nonlinear circuits. IEEE Trans. Circuits Syst.–I: Fund. Th. Appl. **40**, 872–877 (1993)
22. Perez-Munuzuri, V., Perez-Villar, V., Chua, L.O.: Autowaves for image processing on a two–dimensional cnn array of excitable nonlinear circuits: flat and wrinkled labyrinths. IEEE Trans. Circuits Syst.–I: Fund. Th. Appl. **40**, 174–181 (1993)
23. Pivka, L.: Autowaves and spatio-temporal chaos in CNNs—Part i: a tutorials. IEEE Trans. Circuits Syst.–I: Fund. Th. Appl. **42**, 638–649 (1995)
24. Ray, A., Saha, D.C., Saha, P., Chowdhury, A.R.: Generation of amplitude death and rhythmogenesis in coupled hidden attractor system with experimental demonstration. Nonlinear Dyn. 1–12 (2016). doi:10.1007/s11071-016-3121-6
25. Theesar, S.J.S., Banerjee, S., Balasubramaniam, P.: Synchronization of chaotic systems under sampled–data control. Nonlinear Dyn. **70**, 1977–1987 (12)
26. Vembarasan, V., Balasubramaniam, P.: Chaotic synchronization of Rikitake system based on T-S fuzzy control techniques. Nonlinear Dyn. **74**, 31–44 (2013)
27. Vincent, U.E.: Synchronization of rikitake chaotic attractor using active control. Phys. Lett. A **343**, 133–138 (2005)
28. Volos, C.K., Kyprianidis, I.M., Stouboulos, I.N.: Anti-phase and inverse π-lag synchronization in coupled Duffing-type circuits. Int. J. Bif. Chaos **21**, 2357–2368 (2011)
29. Volos, C.K., Kyprianidis, I.M., Stouboulos, I.N.: Various synchronization phenomena in bidirectionally coupled double scroll circuits. Commun. Nonlinear Sci. Numer. Simul. **16**, 3356–3366 (2011)
30. Wu, C.W.: Synchronization in Coupled Chaotic Circuits and System, 1st edn. World Scientific, Singapore (2002)
31. Wu, X.J., Liu, J.S., Chen, G.R.: Chaos synchronization of Rikitake chaotic attractor using the passive control technique. Nonlinear Dyn. **53**, 45–53 (2008)
32. Yang, T., Chua, L.O.: Control of chaos using sampled-data feedback control. Int. J. Bifuric. Chaos **8**, 2433–2438 (1998)
33. Yassen, M.T.: Chaos control of chaotic dynamical systems using backstepping design. Chaos Solitions Fractals **27**, 537–548 (2006)
34. Yau, H.T., Yan, J.J.: Design of sliding mode controller for lorenz chaotic system with nonlinear input. Chaos Solitions Fractals **19**, 891–898 (2004)
35. Zhang, H., Liu, D., Wang, Z.: Controlling Chaos: Suppression. Synchronization and Chaotification. Springer, Germany (2009)

Chapter 6
Circuitry Realization

6.1 Basic Electronic Components and Electronic Circuits

Using electronic components to realize theoretical models has received significant attention in the literature [2, 7]. From the practical point of view of applications, the implementation of electronic circuits based on mathematical models is a vital topic. On the one hand, the hardware implementation is an effective approach to investigate theoretical models besides the conventional computer-based approach. There are a number of important differences between circuit-based approach and computer-based approach. Firstly, by using electronic circuit we reduce the uncertainties because of systematic and statistical errors in numerical simulations [3]. For instance, when running computer-based simulations we must consider the discretization and round-off errors in the numerical procedures or finite-time approximation of a quantity which is properly described by an infinite-time integral. Secondly, we can display and observe the signals of electronic oscillators on the oscilloscope conveniently. In some cases, the time of numerical simulations is long, whereas running time of circuit is relatively short. Furthermore, the experimental bifurcation diagram can be obtained easily by varying the value of circuital components such as resistor. Therefore, a wide range of dynamical behavior of a circuit can be compared to the numerical simulation results of the corresponding theoretical model. On the other hand, realized electronic circuits can be used directly in numerous practical applications such as secure communications [1], target detection [6], robotics [4], random generator [8], and image encryption process [4, 5].

In this work, in order to realize systems with hidden attractors, we have used only basic electronic components, such as resistors, capacitors, diodes, operational amplifiers, and analog multipliers, as illustrated in Fig. 6.1. Based on such basic electronic components, the general building blocks for nonlinear circuits are designed. Four fundamental blocks (inverting amplifier, op-amp integrator, analog multiplier, and absolute-function circuit) are presented in Fig. 6.2.

An inverting amplifier includes an operational amplifier and two resistors (see Fig. 6.2a). The relationship between the input and the output is described by

© The Author(s) 2017
V.-T. Pham et al., *Systems with Hidden Attractors*,
SpringerBriefs in Nonlinear Circuits, DOI 10.1007/978-3-319-53721-4_6

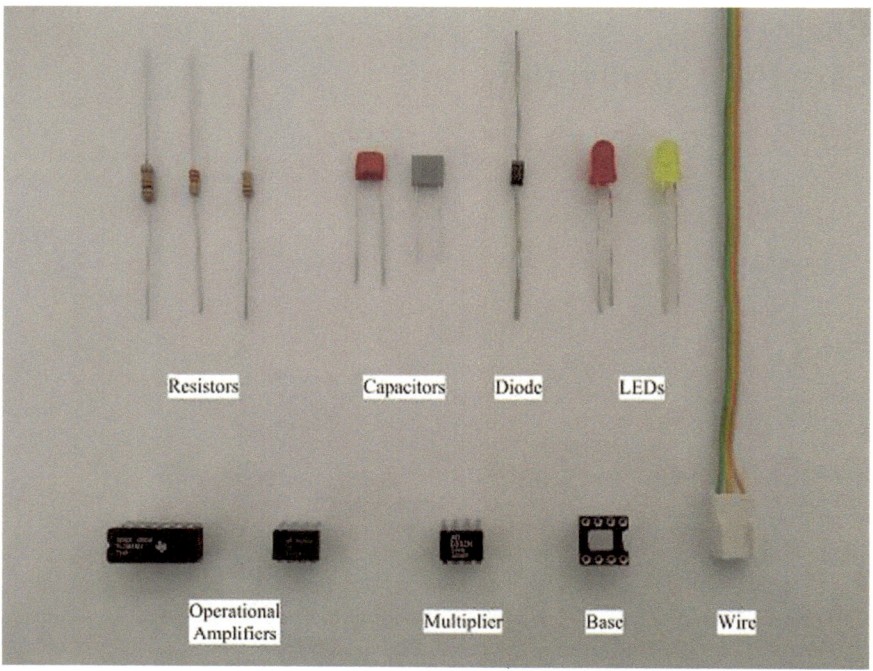

Fig. 6.1 Basic electronic components for realizing electronic circuit: resistors, capacitors, diode, LEDs (in the first row) and operational amplifiers, analog multiplier, base, wire (in the second row)

$$V_o = -\frac{R_2}{R_1}V_i \tag{6.1}$$

Obviously, the gain of the inverting amplifier is defined by the ratio of R_2 to R_1.

An operational amplifier circuit that performs the mathematical operation of integration is presented in Fig. 6.2b. It is trivial to verify that the op-amp integrator produces an output voltage which is proportional to the integral of the input voltage:

$$V_o = -\frac{1}{RC}\int_0^t V_i dt \tag{6.2}$$

As can be seen in Fig. 6.2c, the analog multiplier such as AD633 is connected with two additional resistors. The transfer function for the multiplier is

$$V_o = \frac{\left(V_{iX_1} - V_{iX_2}\right)\left(V_{iY_1} - V_{iY_2}\right)}{10V}\frac{R_1 + R_2}{R_1} \tag{6.3}$$

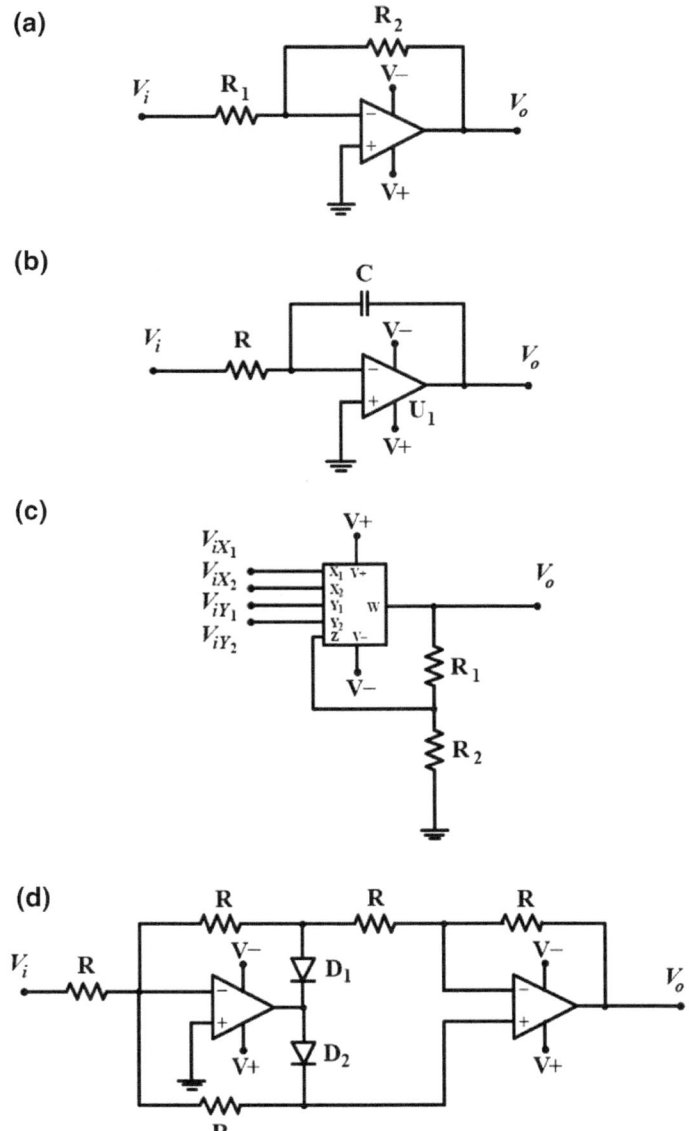

Fig. 6.2 Fundamental functional blocks which are implemented with basic electronic components, **a** inverting amplifier, **b** op-amp integrator, **c** analog multiplier, **d** absolute-function circuit

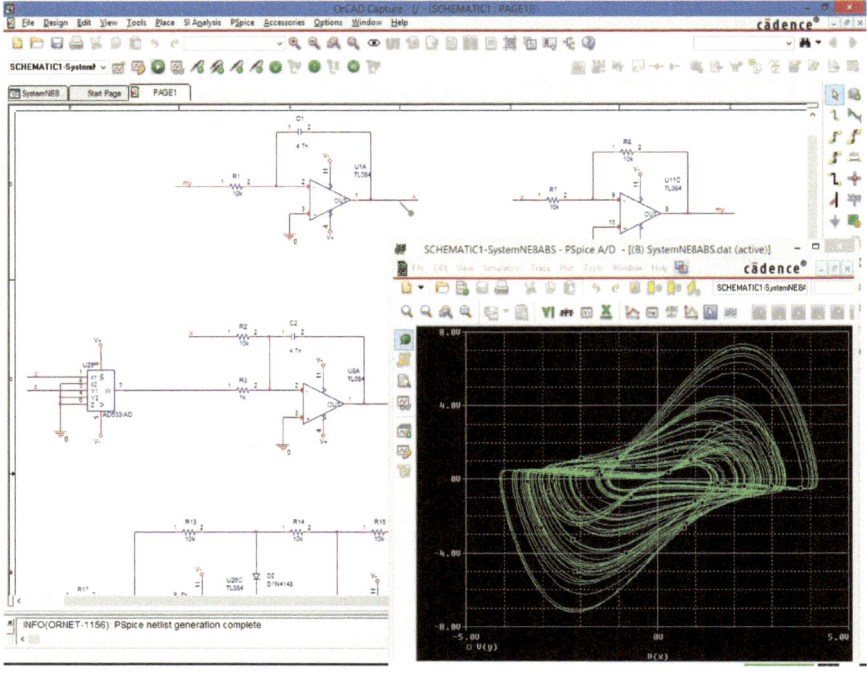

Fig. 6.3 Implementation of a circuit in a circuit simulator and its PSpice result

In a more common case, we do not use the two resistors (R_1 and R_2), and the Z is grounded. The input–output characteristic (6.3) is simplified as follows:

$$V_o = \frac{\left(V_{iX_1} - V_{iX_2}\right)\left(V_{iY_1} - V_{iY_2}\right)}{10V} \tag{6.4}$$

The circuitry for implementing the absolute nonlinearity is presented in Fig. 6.2d. The circuitry is based on two operational amplifiers and two diodes. From Fig. 6.2d, the relationship between the output and the input is derived as follows:

$$V_o = |V_i| \tag{6.5}$$

In general, there are three main steps to realize a system. Firstly, based on the theoretical model, the corresponding electronic circuit is designed with the general building blocks. In the next step, the circuit is implemented in circuit simulators as illustrated in Fig. 6.3. Then, the circuit is realized by using off-the-shelf electronic components. After realizing the circuit, signals of the real circuits are measured and displayed on an oscilloscope (see Fig. 6.4).

Fig. 6.4 Measuring signals from the implemented electronic circuit

6.2 Circuit Implementation of a System with One Stable Equilibrium

In this section, we present an electronic circuit for emulating Wang–Chen system (2.2). As have been presented, system (2.2) has only one stable equilibrium. The schematic of the circuit is shown in Fig. 6.5. The circuit includes three integrators (U_1–U_3), two analog multipliers (U_3 and U_4), eight resistors, and three capacitors. Its experimental realization is presented in Fig. 6.6. The values of electronic component are selected as follows:

$$R_1 = 1\,k\Omega,\ R = R_2 = R_3 = R_4 = R_5 = R_7 = 10\,k\Omega,\ R_6 = 2.5\,k\Omega,\ R_8 = 90\,k\Omega$$
$$V_a = -6m\,V_{DC},\ V_1 = -1\,V_{DC},\ C_1 = C_2 = C_3 = C = 10\,nF$$

$$(6.6)$$

The circuital equations of the circuit are given as follows:

$$\begin{cases} \frac{dv_{C_1}}{dt} = \frac{1}{RC_1}\left(\frac{R}{R_1 10V}v_{C_2}v_{C_3} - \frac{R}{R_2}V_a\right) \\ \frac{dv_{C_2}}{dt} = \frac{1}{RC_2}\left(\frac{R}{R_3 10V}\frac{R_7+R_8}{R_7}v_{C_1}^2 - \frac{R}{R_4}v_{C_2}\right) \\ \frac{dv_{C_3}}{dt} = \frac{1}{RC_3}\left(-\frac{R}{R_5}V_1 - \frac{R}{R_6}v_{C_1}\right) \end{cases}$$

$$(6.7)$$

in which v_{C_1}, v_{C_2}, and v_{C_3} are the voltages across the capacitors C_1, C_2, and C_3.

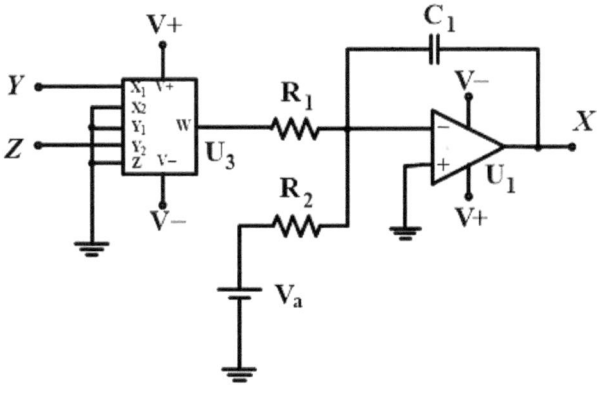

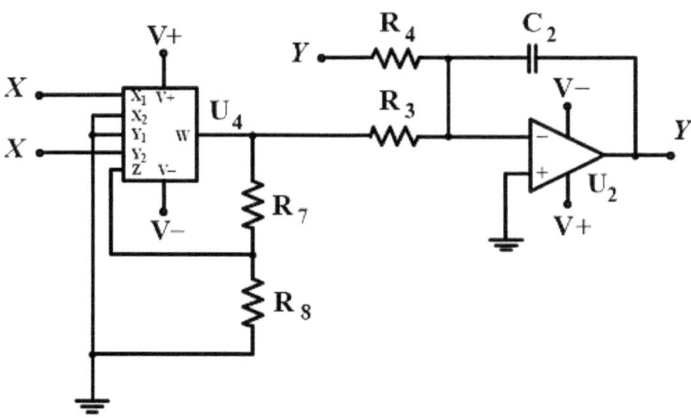

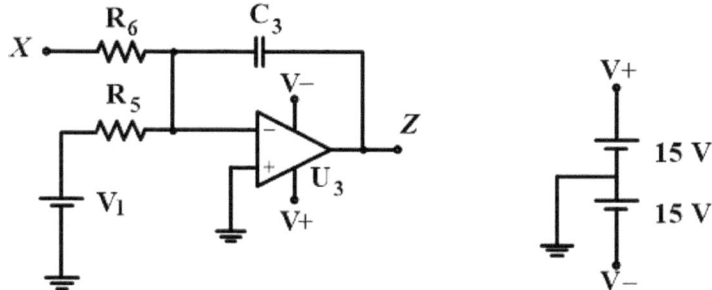

Fig. 6.5 Schematic of the designed circuit which emulates Wang–Chen system with only one stable equilibrium (2.2). Here, the capacitors are $C_1 = C_2 = C_3 = C$

Fig. 6.6 Experimental realization of Wang–Chen system with only one stable equilibrium (2.2)

By normalizing the system (6.7) with $\tau = \frac{t}{RC}$, we have

$$\begin{cases} \dot{X} = \frac{R}{R_1 10} YZ - \frac{R}{R_2 1V} V_a \\ \dot{Y} = \frac{R}{R_3 10} \frac{R_7+R_8}{R_7} X^2 - \frac{R}{R_4} Y \\ \dot{Z} = -\frac{R}{R_5 1V} V_1 - \frac{R}{R_6} X \end{cases} \qquad (6.8)$$

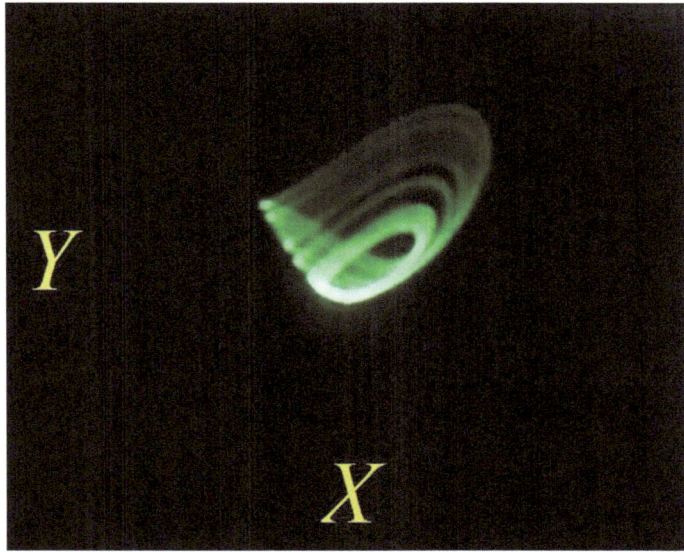

Fig. 6.7 Experimental chaotic phase portrait in $X - Y$ plane of the circuit emulating Wang–Chen system without equilibrium

in which $X = v_{C_1}$, $Y = v_{C_2}$, and $Z = v_{C_3}$.

It is easy to see that the system (6.8) corresponds to the system (2.2) with

$$a = -\frac{R}{R_2}\frac{V_a}{1V} \tag{6.9}$$

For the selected set of components (6.6), the parameters in the system (2.2) are

$$a = 0.006 \tag{6.10}$$

Experimental phase portrait is captured from the oscilloscope as shown in Fig. 6.7.

6.3 Circuit Implementations of Systems with Infinite Equilibria

6.3.1 Circuit Implementation of a System with Line Equilibrium

Electronic circuit for realizing the system with open curve equilibrium (3.14) is presented in this section. In order to avoid problems in the system's realization, the state variables (x, y, z) of the system are scaled up with a factor 10. As a result, the system (3.14) is rewritten as:

$$\begin{cases} \dot{X} = Y \\ \dot{Y} = -X + \frac{1}{10}YZ \\ \dot{Z} = a\,|X| - \frac{b}{10}XY - \frac{1}{10}XZ \end{cases} \tag{6.11}$$

in which $X = 10x$, $Y = 10y$, $Z = 10z$.

The circuit is designed and presented in Fig. 6.8. Its experimental realization is presented in Fig. 6.9. The values of electronic component are selected as:

$$R = 22\,\text{k}\Omega, \quad R_a = 78\,\text{k}\Omega, \quad R_b = 1.158\,\text{k}\Omega, \quad R_1 = 10\,\text{k}\Omega, \\ C_1 = C_2 = C_3 = C = 4.7\,\text{nF} \tag{6.12}$$

The circuital equations of the circuit are given as

$$\begin{cases} \frac{dv_{C_1}}{dt} = \frac{1}{RC_1}\left(v_{C_2}\right) \\ \frac{dv_{C_2}}{dt} = \frac{1}{RC_2}\left(-v_{C_1} + \frac{1}{10V}v_{C_2}v_{C_3}\right) \\ \frac{dv_{C_3}}{dt} = \frac{1}{RC_3}\left(\frac{R}{R_a}\,|v_{C_1}| - \frac{R}{R_b 10V}v_{C_1}v_{C_2} - \frac{1}{10V}v_{C_1}v_{C_3}\right) \end{cases} \tag{6.13}$$

in which v_{C_1}, v_{C_2}, v_{C_3} are the voltages across the capacitors C_1, C_2, and C_3.

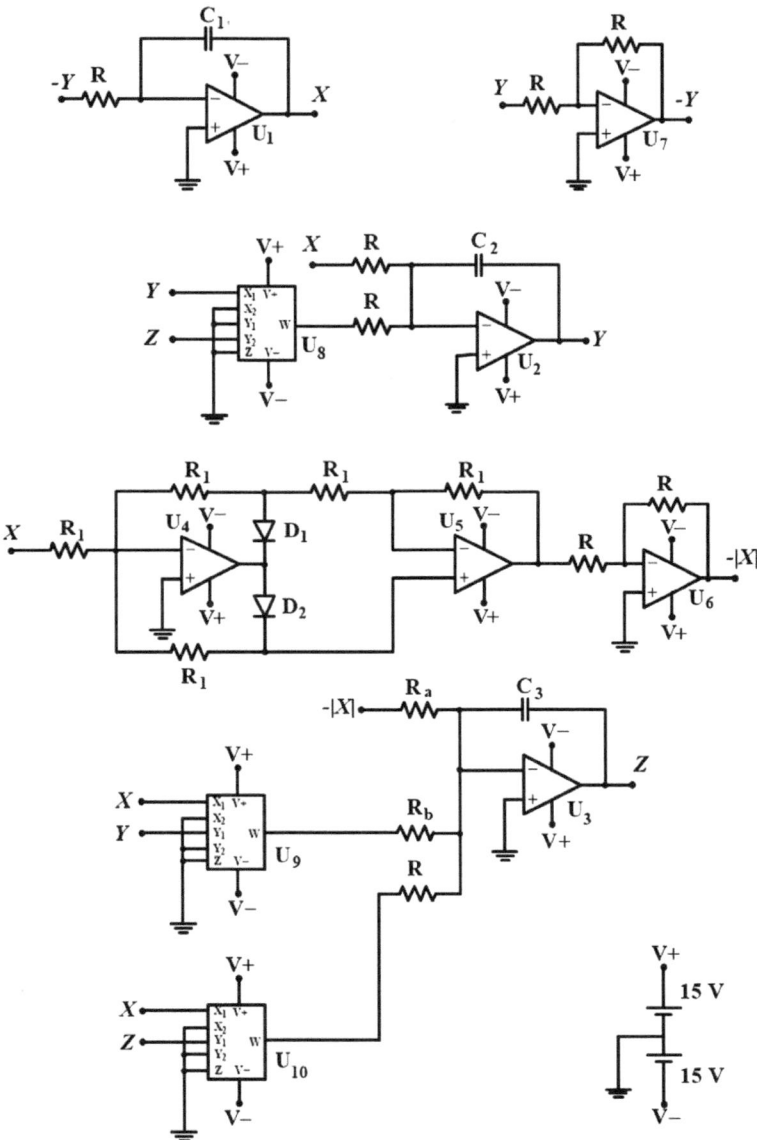

Fig. 6.8 Schematic of the designed circuit which emulates the system with a line equilibrium (6.11). The circuit includes three integrators (U_1–U_3) and three analog multipliers (U_8–U_{10}). Here, the capacitors are $C_1 = C_2 = C_3 = C$

Fig. 6.9 Experimental realization of the proposed system the system with a line equilibrium (6.23)

By normalizing the system (6.13) with $\tau = \frac{t}{RC}$, we have

$$\begin{cases} \dot{X} = Y \\ \dot{Y} = -X + \frac{1}{10}YZ \\ \dot{Z} = \frac{R}{R_a}|X| - \frac{R}{R_b 10}XY - \frac{1}{10}XZ \end{cases} \qquad (6.14)$$

in which $X = v_{C_1}$, $Y = v_{C_2}$, and $Z = v_{C_3}$.

It is easy to see that the system (6.14) corresponds to the system (6.11) with

$$a = \frac{R}{R_a}, \quad b = \frac{R}{R_b} \qquad (6.15)$$

For the selected set of components (6.12), the parameters in the system (6.11) are

$$a = 0.25, \ b = 19 \qquad (6.16)$$

Experimental phase portrait is captured from the oscilloscope as shown in Fig. 6.10.

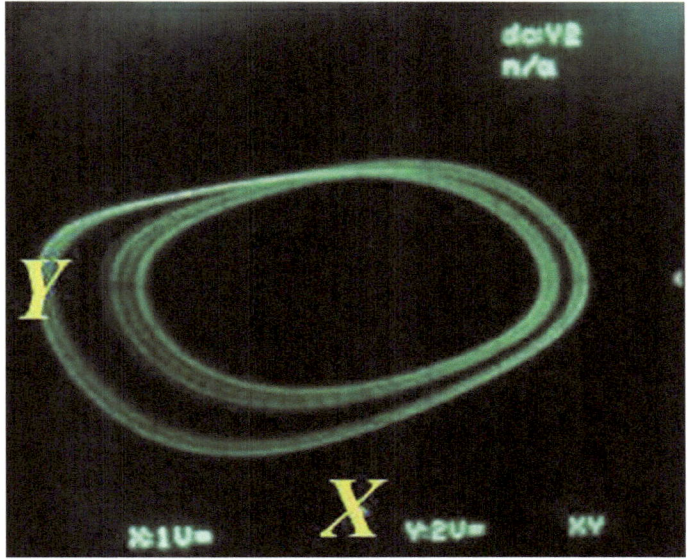

Fig. 6.10 Experimental phase portrait in $X - Y$ plane of the circuit with a line of equilibrium points

6.3.2 Circuit Implementation of a System with Closed Curve Equilibrium

In this section, we present an electronic circuit for emulating the mathematical system (3.34). In order to avoid problems in the system's realization, the state variables (x, y, z) of the system are scaled up. The system (3.34) is transformed to

$$
\begin{cases}
\dot{X} = aZ \\
\dot{Y} = \frac{b}{10}XZ + \frac{c}{10^2}Z^3 \\
\dot{Z} = \frac{1}{10^3}X^4 + \frac{1}{10^3}Y^4 + \frac{d}{10}XZ - 10r^2
\end{cases}
\tag{6.17}
$$

in which $X = 10x$, $Y = 10y$, and $Z = 10z$.

The circuit is designed and presented in Fig. 6.11. The circuit includes three integrators (U_1–U_3) and eight analog multipliers (U_4–U_{11}). Its experimental realization is presented in Fig. 6.12. The values of electronic component are selected as:

$R = 10\,\text{k}\Omega$, $R_a = 100\,\text{k}\Omega$, $R_b = 3.3\,\text{k}\Omega$, $R_c = 4.545\,\text{k}\Omega$, $R_d = 50\,\text{k}\Omega$, $R_r = 1.234\,\text{k}\Omega$
$V_r = 1V_{DC}$, $C_1 = C_2 = C_3 = C = 10\,\text{nF}$

$$\tag{6.18}$$

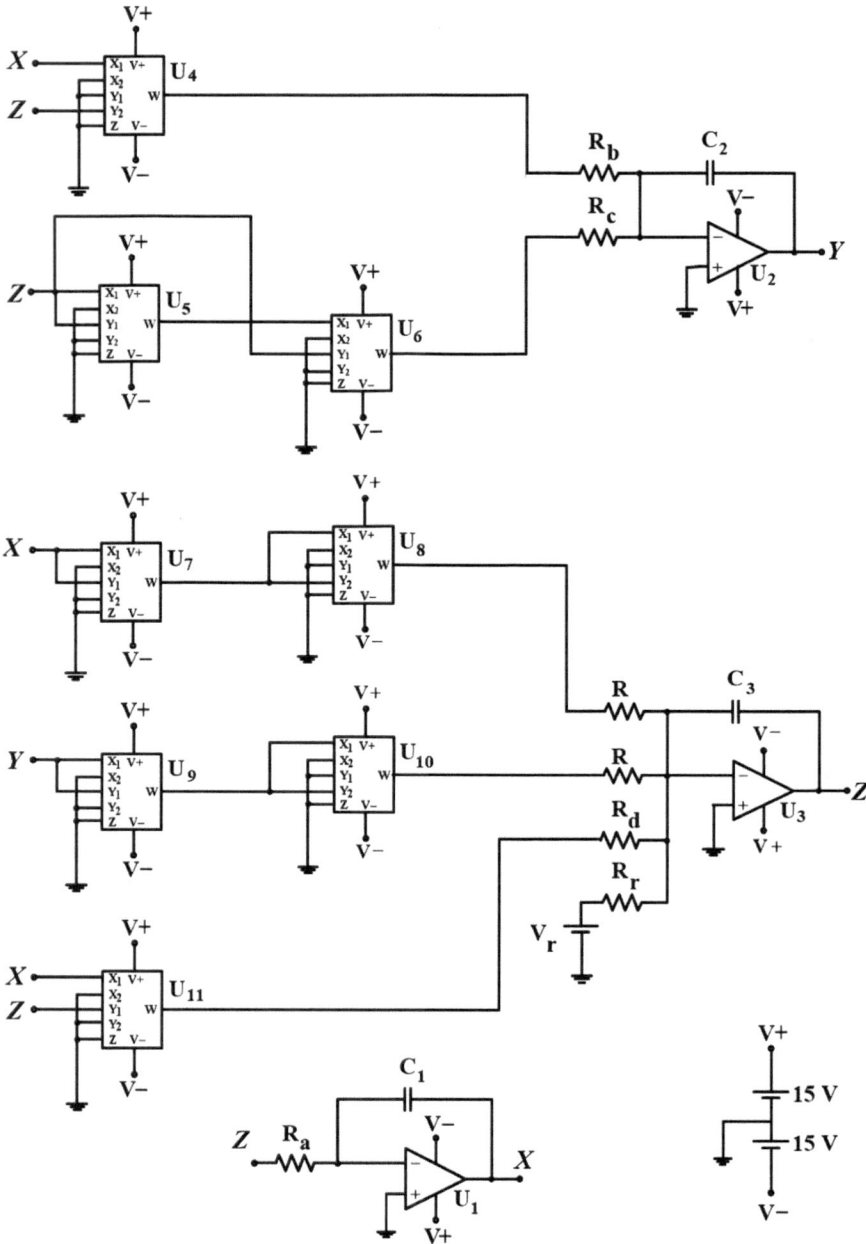

Fig. 6.11 Schematic of the designed circuit which emulates the system with infinite equilibria (6.17). Here, the capacitors are $C_1 = C_2 = C_3 = C$

Fig. 6.12 Experimental realization of the proposed system with infinite equilibria (6.17)

The circuital equations of the circuit are given as:

$$
\begin{cases}
\frac{dv_{C_1}}{dt} = \frac{1}{RC_1}\left(-\frac{R}{R_a}v_{C_3}\right) \\
\frac{dv_{C_2}}{dt} = \frac{1}{RC_2}\left(\frac{R}{R_b 10V}v_{C_1}v_{C_3} - \frac{R}{R_c 10^2 V^2}v_{C_3}^3\right) \\
\frac{dv_{C_3}}{dt} = \frac{1}{RC_3}\left(\frac{1}{10^3 V^3}v_{C_1}^4 + \frac{1}{10^3 V^3}v_{C_2}^4 - \frac{R}{R_d 10V}v_{C_1}v_{C_3} - \frac{R}{R_r}V_r\right)
\end{cases}
\tag{6.19}
$$

in which v_{C_1}, v_{C_2}, and v_{C_3} are the voltages across the capacitors C_1, C_2, and C_3.

By normalizing the system (6.19) with $\tau = \frac{t}{RC}$, we have

$$
\begin{cases}
\dot{X} = -\frac{R}{R_a}Z \\
\dot{Y} = \frac{R}{R_b 10}XZ - \frac{R}{R_c 10^2}Z^3 \\
\dot{Z} = \frac{1}{10^3}X^4 + \frac{1}{10^3}Y^4 - \frac{R}{R_d 10}XZ - \frac{R}{R_r}V_r
\end{cases}
\tag{6.20}
$$

in which $X = v_{C_1}$, $Y = v_{C_2}$, and $Z = v_{C_3}$.

It is easy to see that the system (6.20) corresponds to the system (6.17) with

$$
a = -\frac{R}{R_a}, \quad b = \frac{R}{R_b}, \quad c = -\frac{R}{R_c}, \quad d = \frac{R}{R_d}, \quad r^2 = \frac{R}{10R_r}V_r
\tag{6.21}
$$

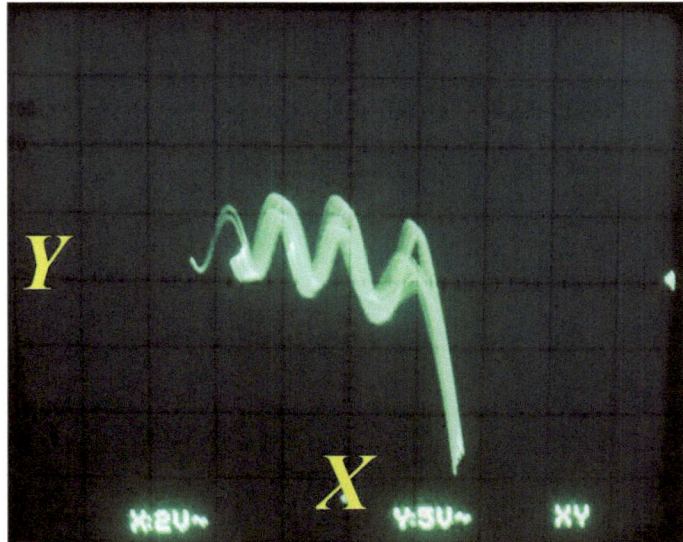

Fig. 6.13 Experimental phase portrait in $X-Y$ plane of the circuit with rounded-square equilibrium

For the selected set of components (6.18), the parameters in the system (6.17) are

$$a = -0.1, \ b = 3, \ c = -2.2, \ d = -0.2, \ r = 0.9 \qquad (6.22)$$

Experimental phase portrait is captured from the oscilloscope as shown in Fig. 6.13.

6.3.3 Circuit Implementation of a System with Open Curve Equilibrium

Electronic circuit for realizing the system with open curve equilibrium (3.41) is discussed in this section. In order to avoid problems in the system's realization, the state variables (x, y, z) of the system are scaled up with a factor 10. As a result, the system (3.41) becomes

$$\begin{cases} \dot{X} = -aZ \\ \dot{Y} = \frac{1}{10^2} X Z^2 \\ \dot{Z} = X - b|Y| + \frac{c}{10^2} Y^2 Z - \frac{1}{10^2} Z^3 \end{cases} \qquad (6.23)$$

in which $X = 10x$, $Y = 10y$, and $Z = 10z$.

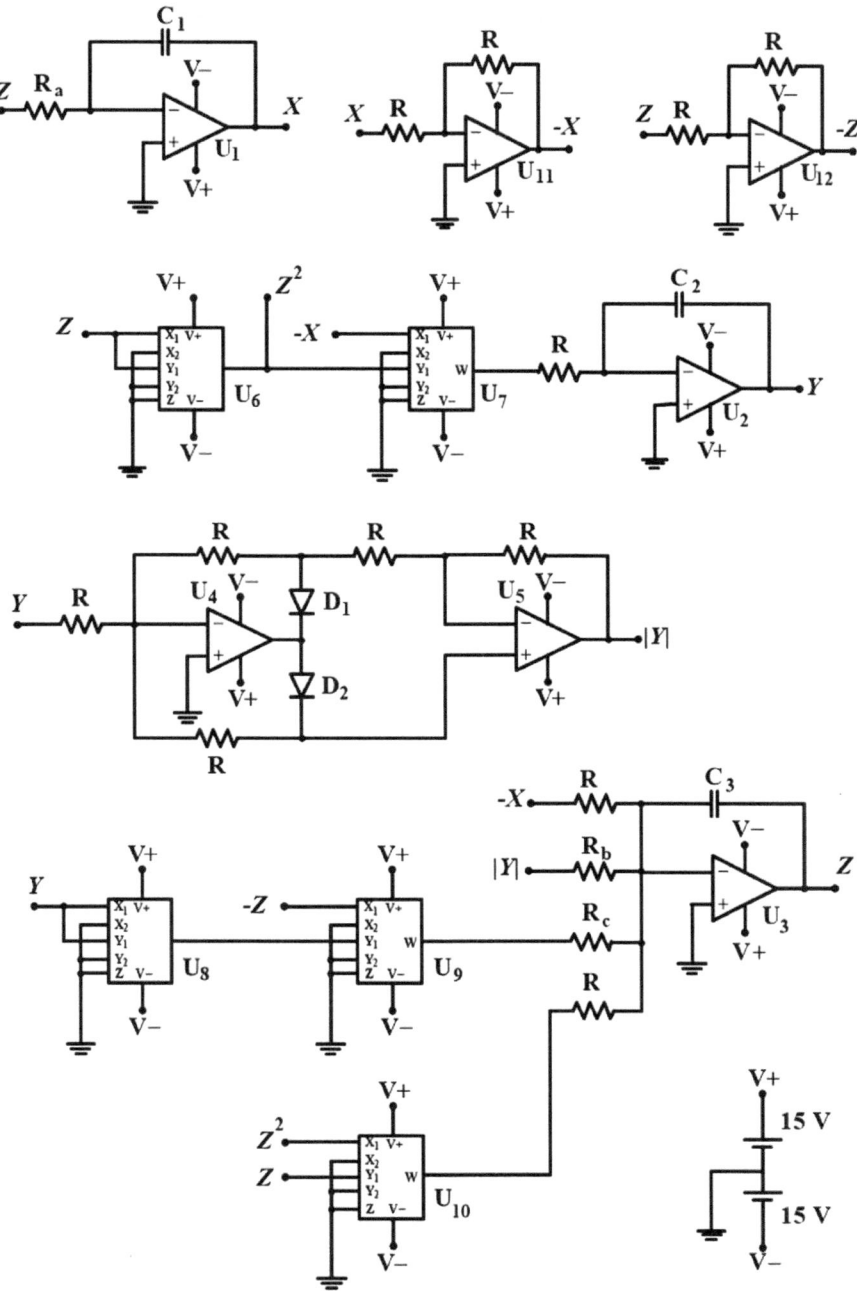

Fig. 6.14 Schematic of the designed circuit which emulates the system with infinite equilibria (6.23). Here the capacitors are $C_1 = C_2 = C_3 = C$

Fig. 6.15 Experimental realization of the proposed system with infinite equilibria (6.23)

The circuit is designed and presented in Fig. 6.14. The circuit includes three integrators (U_1–U_3) and five analog multipliers (U_6–U_{10}). Its experimental realization is presented in Fig. 6.15. The values of electronic component are selected as:

$$R = 10\,\mathrm{k\Omega},\ \ R_a = 10\,\mathrm{k\Omega},\ R_b = 25\,\mathrm{k\Omega},\ \ R_c = 3.333\,\mathrm{k\Omega},$$
$$C_1 = C_2 = C_3 = C = 4.7\,\mathrm{nF} \tag{6.24}$$

The circuital equations of the circuit are given as:

$$\begin{cases} \frac{dv_{C_1}}{dt} = \frac{1}{RC_1}\left(-\frac{R}{R_a}v_{C_3}\right) \\ \frac{dv_{C_2}}{dt} = \frac{1}{RC_2}\left(\frac{1}{10^2 V^2}v_{C_1}v_{C_3}^2\right) \\ \frac{dv_{C_3}}{dt} = \frac{1}{RC_3}\left(v_{C_1} - \frac{R}{R_b}\left|v_{C_2}\right| + \frac{R}{R_c 10^2 V^2}v_{C_2}^2 v_{C_3} - \frac{1}{10^2 V^2}v_{C_3}^3\right) \end{cases} \tag{6.25}$$

in which v_{C_1}, v_{C_2}, and v_{C_3} are the voltages across the capacitors C_1, C_2, and C_3.
By normalizing the system (6.25) with $\tau = \frac{t}{RC}$, we have

$$\begin{cases} \dot{X} = -\frac{R}{R_a}Z \\ \dot{Y} = \frac{1}{10^2}XZ^2 \\ \dot{Z} = X - \frac{R}{R_b}|Y| + \frac{R}{R_c 10^2}Y^2 Z - \frac{1}{10^2}Z^3 \end{cases} \tag{6.26}$$

in which $X = v_{C_1}$, $Y = v_{C_2}$, and $Z = v_{C_3}$.
It is easy to see that the system (6.26) corresponds to the system (6.23) with

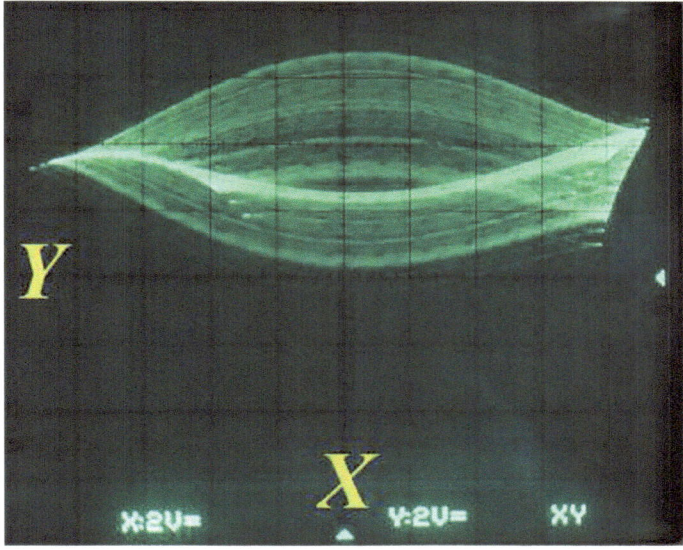

Fig. 6.16 Experimental phase portrait in $X - Y$ plane of the circuit with a piecewise linear curve of equilibrium points

$$a = \frac{R}{R_a}, \; b = \frac{R}{R_b}, \; c = \frac{R}{R_c} \tag{6.27}$$

For the selected set of components (6.24), the parameters in the system (6.23) are

$$a = 1, \; b = 0.4, \; c = 3 \tag{6.28}$$

Experimental phase portrait is captured from the oscilloscope as shown in Fig. 6.16.

6.4 Circuit Implementation of a System Without Equilibrium

In this section, we present an electronic circuit for emulating the mathematical system (4.21). As have been presented, system (4.21) has no equilibrium. The schematic of the circuit is shown in Fig. 6.17. The circuit includes three integrators (U_1–U_3) and two analog multipliers (U_8, U_9). Its experimental realization is presented in Fig. 6.18 The values of electronic components are selected as:

$$\begin{aligned} R &= 10\,\text{k}\Omega, \; V_a = 1.35 V_{DC} \\ C_1 &= C_2 = C_3 = C = 10\,\text{nF} \end{aligned} \tag{6.29}$$

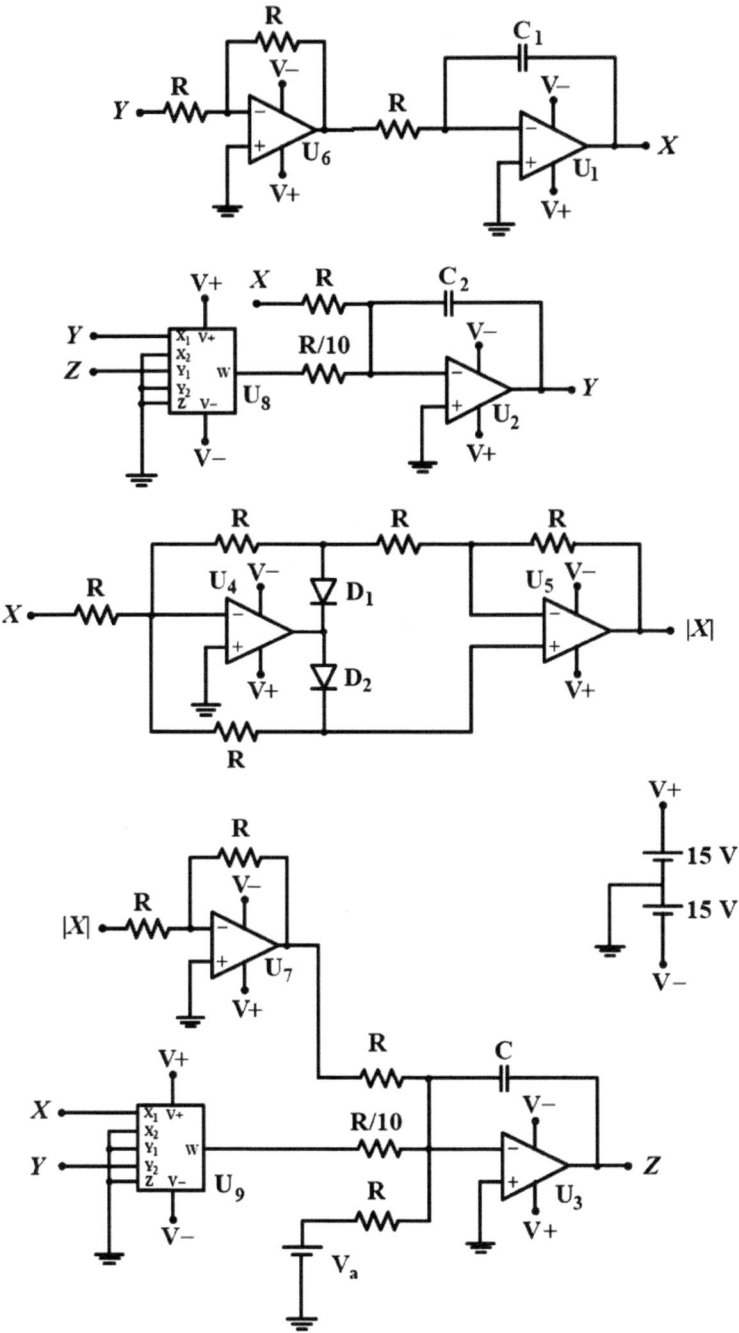

Fig. 6.17 Schematic of the designed circuit which emulates the system without equilibrium (4.21). Here, the capacitors are $C_1 = C_2 = C_3 = C$

Fig. 6.18 Experimental realization of the system without equilibrium (4.21)

The circuital equations of the circuit are given as

$$
\begin{cases}
\frac{dv_{C_1}}{dt} = \frac{1}{RC_1}\left(v_{C_2}\right) \\
\frac{dv_{C_2}}{dt} = \frac{1}{RC_2}\left(-v_{C_1} - \frac{1}{1V}v_{C_2}v_{C_3}\right) \\
\frac{dv_{C_3}}{dt} = \frac{1}{RC_3}\left(|v_{C_1}| + \frac{1}{1V}v_{C_1}v_{C_2} - V_a\right)
\end{cases}
\tag{6.30}
$$

in which v_{C_1}, v_{C_2}, and v_{C_3} are the voltages across the capacitors C_1, C_2, and C_3.
By normalizing the system (6.30) with $\tau = \frac{t}{RC}$, we have

$$
\begin{cases}
\dot{X} = Y \\
\dot{Y} = -X - YZ \\
\dot{Z} = |X| + XY - a
\end{cases}
\tag{6.31}
$$

in which $X = v_{C_1}$, $Y = v_{C_2}$, and $Z = v_{C_3}$.
It is easy to see that the system (6.31) corresponds to the system (4.21) with

$$
a = \frac{V_a}{1V}
\tag{6.32}
$$

For the selected set of components (6.29), the parameter in the system (4.21) is

$$
a = 1.35
\tag{6.33}
$$

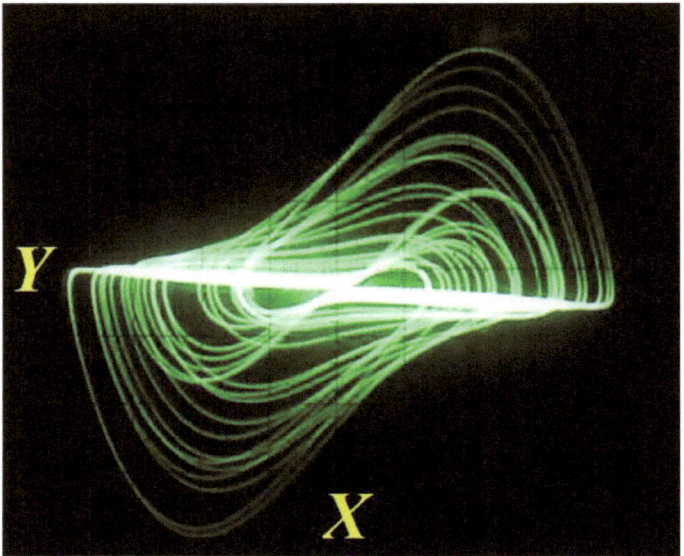

Fig. 6.19 Experimental phase portrait in $X - Y$ plane of the circuit without equilibrium

Experimental phase portrait is captured from the oscilloscope as shown in Fig. 6.19.

6.5 Circuit Implementation of a System with Different Families of Hidden Attractors

Electronic circuit for realizing the system with different families of hidden attractors (4.26) is discussed in this part. In order to avoid problems in the systems realization, the state variables (x, y, z) of the system are scaled up with a factor 5. As a result, system (4.26) becomes

$$\begin{cases} \dot{X} = Y \\ \dot{Y} = 0.08XZ - 5a \\ \dot{Z} = 0.3X - 0.1Z - 0.28Y^2 - \frac{b}{5}XY - 5c \end{cases} \tag{6.34}$$

in which $X = 5x$, $Y = 5y$, $Z = 5z$.

The circuit is designed and presented in Fig. 6.20. Its experimental realization is presented in Fig. 6.21. The values of electronic component are selected as:

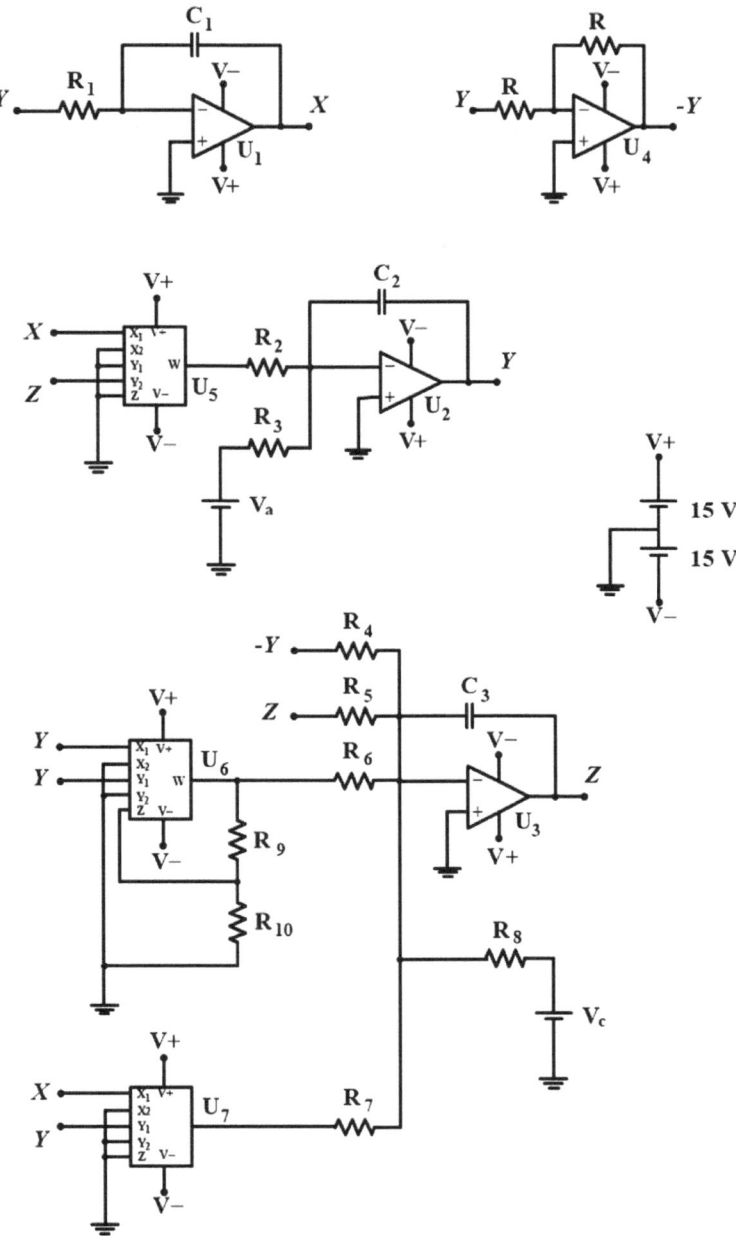

Fig. 6.20 Schematic of the designed circuit which emulates the system with variable equilibria (6.17). The circuit includes three integrators (U_1–U_3) and three analog multipliers (U_5–U_7). Here, the capacitors are $C_1 = C_2 = C_3 = C$

Fig. 6.21 Experimental realization of the proposed system with variable equilibria (6.17)

$$R = R_1 = R_3 = R_8 = R_9 = 10\,\text{k}\Omega, \ \ R_2 = 12.5\,\text{k}\Omega, \ R_4 = 33.333\,\text{k}\Omega,$$
$$R_5 = 100\,\text{k}\Omega, \ \ R_6 = 35.714\,\text{k}\Omega, \ \ R_7 = 25\,\text{k}\Omega, \ \ R_{10} = 90\,\text{k}\Omega, \tag{6.35}$$
$$V_a = V_c = 0V_{DC}, \ \ C_1 = C_2 = C_3 = C = 6.8\,\text{nF}$$

The circuital equations of the circuit are given as:

$$\begin{cases} \frac{dv_{C_1}}{dt} = \frac{1}{RC_1}\left(\frac{R}{R_1}v_{C_2}\right) \\ \frac{dv_{C_2}}{dt} = \frac{1}{RC_2}\left(\frac{R}{R_2 10V}v_{C_1}v_{C_3} - \frac{R}{R_3}V_a\right) \\ \frac{dv_{C_3}}{dt} = \frac{1}{RC_3}\left(\frac{R}{R_4}v_{C_2} - \frac{R}{R_5}v_{C_3} - \frac{R}{R_6 10V}\frac{R_9+R_{10}}{R_9}v_{C_2}^2 - \frac{R}{R_7 10V}v_{C_1}v_{C_2} - \frac{R}{R_8}V_c\right) \end{cases} \tag{6.36}$$

in which v_{C_1}, v_{C_2}, and v_{C_3} are the voltages across the capacitors C_1, C_2, and C_3.

By normalizing the system (6.36) with $\tau = \frac{t}{RC}$, we have

$$\begin{cases} \dot{X} = \frac{R}{R_1}Y \\ \dot{Y} = \frac{R}{R_2 10}XZ - \frac{R}{R_3 1V}V_a \\ \dot{Z} = \frac{R}{R_4}Y - \frac{R}{R_5}Z - \frac{R}{R_6 10}\frac{R_9+R_{10}}{R_9}Y^2 - \frac{R}{R_7 10}XY - \frac{R}{R_8 1V}V_c \end{cases} \tag{6.37}$$

in which $X = v_{C_1}$, $Y = v_{C_2}$, and $Z = v_{C_3}$.

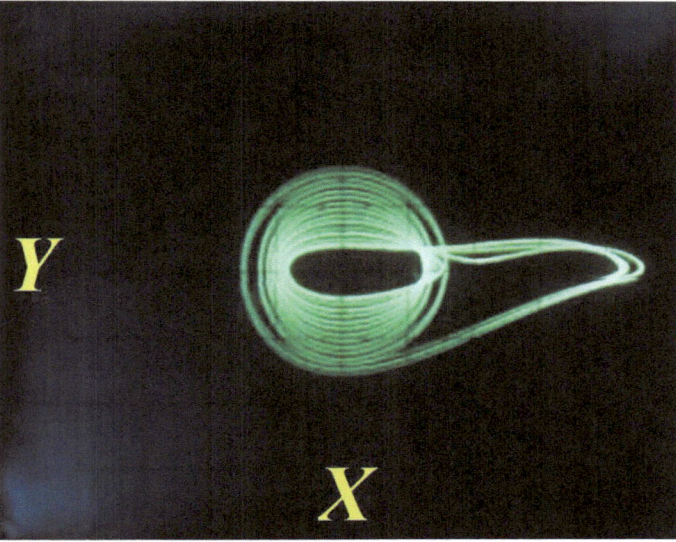

Fig. 6.22 Experimental phase portrait in $X - Y$ plane of the circuit with variable equilibrium

It is easy to see that the system (6.37) corresponds to the system (6.34) with

$$a = \frac{R}{5R_3 1V} V_a, \ b = \frac{R}{2R_7}, \ c = \frac{R}{5R_8 1V} V_c \qquad (6.38)$$

For the selected set of components (6.35), the parameters in the system (6.34) are

$$a = 0, \ b = 0.2, \ c = 0 \qquad (6.39)$$

Experimental phase portrait is captured from the oscilloscope as shown in Fig. 6.22.

References

1. Banerjee, S.: Chaos Synchronization and Cryptography for Secure Communication. IGI Global, USA (2010)
2. Cicek, S., Uyaroglu, Y., Pehlivan, I.: Simulation and circuit implementation of Sprott case H chaotic system and its synchronization application for secure communication systems. J. Circuit Syst. Comp. **22**(1350), 022–15 (2013)
3. Sprott, C.: A proposed standard for the publication of new chaotic systems. Int. J. Bifurc. Chaos **21**, 2391–2394 (2011)
4. Volos, C.K., Kyprianidis, I.M., Stouboulos, I.N.: A chaotic path planning generator for autonomous mobile robots. Robot. Auto. Syst. **60**, 651–656 (2012)

5. Volos, C.K., Kyprianidis, I.M., Stouboulos, I.N.: Image encryption process based on chaotic synchronization phenomena. Signal Process. **93**, 1328–1340 (2013)
6. Wang, B., Xu, H., Yang, P., Liu, L., Li, J.: Target detection and ranging through lossy media using chaotic radar. Entropy **17**, 2082–2093 (2015)
7. Wu, X., He, Y., Yu, W., Yin, B.: A new chaotic attractor and its synchronization implementation. Circuits Syst. Signal Process **34**, 1747–1768 (2015)
8. Yalcin, M.E., Suykens, J.A.K., Vandewalle, J.: True random bit generation from a double–scroll attractor. IEEE Trans. Circuits Syst. I, Regular Papers **51**, 1395–1404 (2004)

Chapter 7
Concluding Remarks

This book provides a brief summarization of systems with hidden attractors. We have first presented the hidden attractor and its presence in different systems. Three main families of systems with hidden attractors, systems with stable equilibrium, systems with an infinite number of equilibrium points, and systems without equilibrium, have been introduced. Mathematical models of systems with hidden attractors are reported. We have then investigated the synchronization of systems with hidden attractors via two approaches—diffusion coupling and nonlinear control. We have discussed a design procedure to realize such systems with electronic components. In addition, we have also illustrated the procedure through various examples of electronic circuits. Experimental results show the feasibility of theoretical systems with hidden attractors. It is noteworthy that circuits are implemented with common off-the-shelf electronic components. Furthermore, experiments are easy to be performed in a laboratory. In order to display experimental results, it just needs an oscilloscope.

In recent years, systems with hidden attractors have received great attention from both theoretical and practical viewpoints. There are a number of important differences between self-excited attractors and hidden attractors. Self-excited attractor can be localized straightforwardly by applying a standard computational procedure. By contrast, we have to develop a specific computational procedure to identify a hidden attractor due to the fact that the equilibrium points do not help in their localization. There is evidence that hidden attractors play a crucial role in the fields of oscillators, describing convective fluid motion, model of drilling system, or multilevel DC/DC converter. In addition, hidden attractors are attracting widespread interest because they may lead to unexpected and disastrous responses, for example in a structure like a bridge or an airplane wing. Therefore, it is useful for engineering students and researchers to know the emerging topics in this new classification of attractors.

For the past five years, although there has been a rapid rise in the discovery of systems with hidden attractors, there is still very little scientific understanding of hidden attractors. For example, to date there has been little discussion on the existence of systems with different families of hidden attractors [2] or the transformation from a known system to a new system with hidden attractors [1]. Further studies need to be carried out in order to provide insights into hidden attractors.

© The Author(s) 2017
V.-T. Pham et al., *Systems with Hidden Attractors*,
SpringerBriefs in Nonlinear Circuits, DOI 10.1007/978-3-319-53721-4_7

References

1. Pham, V.T., Jafari, S., Kapitaniak, T.: Constructing a chaotic system with an infinite number of equilibrium points. Int. J. Bifurc. Chaos **26**, 1650, 225 (2016)
2. Pham, V.T., Volos, C., Jafari, S., Vaidyanathan, S., Kapitaniak, T., Wang, X.: A chaotic system with different families if hidden attractors. Int. J. Bifurc. Chaos **26**, 1650, 139 (2016)

Index

© The Author(s) 2017
V.-T. Pham et al., *Systems with Hidden Attractors*,
SpringerBriefs in Nonlinear Circuits, DOI 10.1007/978-3-319-53721-4